Is Science Western?

Gregory W. Dawes

Is Science Western?

An Inquiry

palgrave
macmillan

Gregory W. Dawes
Department of Philosophy
University of Otago
Dunedin, New Zealand

ISBN 978-3-032-10640-7 ISBN 978-3-032-10641-4 (eBook)
https://doi.org/10.1007/978-3-032-10641-4

This Palgrave Macmillan imprint is published by the registered company Springer Nature Switzerland AG.
The registered company address is: Gewerbestrasse 11, 6330 Cham, Switzerland

Competing Interests The author has no competing interests to declare that are relevant to the content of this manuscript.

Contents

About the Author

Gregory W. Dawes gained his first graduate degree at the Pontifical Biblical Institute in Rome (1988) before completing PhDs in both Biblical Studies (1995) and Philosophy (2007). His early work examined the challenges to belief that arose from the historical study of Christian origins. This led him to the study of the relations between science and religion and, more generally, of the various 'modes of thought' found across cultures. His books include *Galileo and the Conflict Between Religion and Science* (2016), *Religion, Philosophy and Knowledge* (2016), and *Deprovincializing Science and Religion* (2021), the last of which is devoted to broadening the scope of the philosophical discussion of religion and science.

CHAPTER 1

What Is the Question?

Abstract The word 'science' has three possible uses: a first in which it is synonymous with 'knowledge', a second in which it means any systematized body of knowledge, and a third referring to a particular set of knowledge-related practices those arose in early modern Europe. What could it mean to describe the last of these—modern science—as 'Western'? A first possible meaning is geographical, suggesting that science is practised in 'Western' countries. A second is historical, holding that modern science is Western insofar as it began in Europe. A third is cultural, holding that modern science has features not shared by knowledge-related practices elsewhere. A fourth possible meaning of the phrase is a 'genus and species' one: distinguishing differing forms of science from one another.

Keywords Decolonization • Historicism • Incommensurability • Relativism • Romanticism • Universalism

G. W. Dawes, *Is Science Western?*,
https://doi.org/10.1007/978-3-032-10641-4_1

The question I am addressing arises from a usage that has become common in the humanities: the practice of labelling modern science as 'Western'. A recent Google Scholar search, for example, yielded more than 127,000 instances of the phrase 'Western science'. This way of speaking is particularly common among writers in postcolonial studies, many of whom speak of the need to 'decolonize epistemology' (the theory of knowledge) (Mignolo 2012: 20). Advocates of decolonization commonly speak of 'knowledges' in the plural, often referring to differing 'knowledge systems' (Barnhardt and Kawagley 2005: 9). They argue that different societies have differing 'ways of knowing', 'Western science' being merely one among many. Some such writers go further, claiming that Western science cannot claim to be in any way superior to other ways of knowing. This idea is often buttressed by the claim that these differing forms of knowledge are incommensurable: they cannot be directly compared (Durie 2004: 1138).

The claim that differing knowledge systems are incommensurable is part of a wider tendency to reject claims to universality (Waters 2001: 145). Postcolonial writers regard the idea that science has universal applicability—that it is a form of knowledge for all peoples—as a colonial imposition, whose effect is to marginalize the ways of knowing found in indigenous societies (Seuffert 1997: 98). Rather than being universally applicable, such writers argue, modern science is itself a form of 'local knowledge' (Canagarajah 2002: 245), 'the ethnoscience of the West' (Watson-Verran and Turnbull 1995: 115). I shall refer to this as the 'relativizing' view of science.

The historian cannot help but note a certain irony here. It consists in the fact that the relativizing view has itself emerged from a distinctively European tradition of thought. In its suspicion of claims to universality it follows in the footsteps of the nineteenth-century German Romantics and, before them, Johann Gottfried Herder (1744–1803). Herder argued that each nation has its distinctive way of thinking—its *Nationalgeist*—and vigorously rejected the idea that 'one single culture ... is the epitome of civilization' (Sikka 2005: 312). Herder's criticisms of European imperialism are as forceful as those of any of today's postcolonial writers (Beiser 1992: 203). But while Herder insists on a plurality of culturally distinctive ways of knowing, it is less clear that he thought they were incommensurable. Despite what Isaiah Berlin suggests (2013: 264), Herder seems to have believed that there are universal human needs (and corresponding goods) that can establish culture-transcending standards of judgements (Sikka 2005: 331).

Romanticism contributed to the rise of a second movement in whose footsteps the relativizers are following: that of historicism. The word 'historicism' is used in a bewildering variety of ways (Leerssen 2004: 119). But in one of its uses it refers to the recognition that all human values, institutions, and practices are contingent, having emerged from particular social and political contexts. Long before the rise of postcolonial thinking, writers in the historicist tradition were arguing that the 'scientific world-picture is itself only one among many and, like all the others, it has been produced by a certain society under definite conditions' (Landgrebe 1940: 44). The challenge of historicist thinking went much deeper than a challenge to the universality of science. It called into question all 'moral, political, and religious beliefs and practices', suggesting that had no 'purpose, meaning and validity' beyond the culture in which they arose (Beiser 2011: 11).

Those who oppose the relativizing view of science—I shall call them 'universalists'—do not deny the insights of historicist thinking. They do not deny, for instance, that scientific theories reflect the social and political contexts from which they emerge. If we ask, for example, why Darwin's theory of natural selection emerged in Great Britain (rather than elsewhere), the answer may need to appeal to the competitive ethos of nineteenth-century British economic activity (Greene 1981: 7). But universalists argue that a recognition of the historical character of modern science does not rule out the idea that its results are applicable everywhere (Ryn 2003: 63, 119). A scientific theory may be applicable in every cultural setting, even if it originally emerged from a particular society, and was shaped by its concerns. Although Darwin's theory did owe much to the social world in which it emerged, it still had—universalists would argue—greater explanatory power than any of its rivals.

Which side of this debate has the better arguments? Even if science happens to have developed in 'the West', is it *merely* Western, a local rather than a universal form of knowledge? Does it have distinctively Western characteristics, or are its methods, procedures, and assumptions more widely shared? Those are the questions I shall be addressing.

1.1 The Term 'Science'

Let me begin with some distinctions, for the two terms being used in this debate—'science' and 'Western'—have a range of possible meanings. I shall start with the various possible uses of the word 'science' (and cognate terms in other European languages). Following Theophilus Okere (2005), I shall begin with the broadest use of the term, examining progressively narrower senses, until I arrive at the family of practices I am referring to as 'modern science'.

1.1.1 Science as Knowledge

The broadest use of the word 'science' employs the term as a synonym for 'knowledge'. In English, this usage is now archaic, although it can still be heard in performances of Shakespeare's *Measure for Measure.* In Act 1, Scene 2, Vincentio remarks to Escalus: 'your own science exceeds, in that, the lists of all advice my strength can give you' (I.i.5). In this context, 'science' means simply 'knowledge' or 'learning'. The same use is found in other European languages. One can, for instance, speak in Italian of *la scienza del bene e del male* for 'the knowledge of good and evil', and in French of *la science infuse* for 'innate knowledge' (a phrase used ironically for a 'know-it-all'). All human beings (and arguably many kinds of non-human animals) have 'science' in this sense, that is to say, knowledge in general.

There is, however, a qualification that needs to be made here, for the word 'knowledge' itself has two uses: one descriptive and the other normative. In the descriptive use—common among social scientists—'knowledge' refers to the 'collective beliefs' of a particular community, no distinction being made between true and false beliefs (Pelto and Pelto 1997: 149). In its normative use, on the other hand, 'knowledge' refers to beliefs that emerge from practices that are 'truth-tracking' (Nozick 1981: 172). Philosophers customarily use the term in its normative sense: they do not regard false beliefs as instances of knowledge. Take, for example, the idea that the earth is stationary while the sun moves, a belief held by practically every astronomer before the work of Nicolaus Copernicus (1473–1543). During that period, it could be described as having been knowledge in the descriptive sense. But since it is a false belief, it would not count as 'knowledge' in the normative sense.

I shall not, however, dwell on this distinction here. If we understand 'science' in the broadest sense of the word, as knowledge or learning, then we may safely assume that all peoples have had science, even in the normative sense of the term. They must have had at least some well-founded true beliefs. Without, for example, well-founded beliefs regarding their physical environment, it seems unlikely that any people could have survived for any length of time. Later in this study I shall look more closely at what kinds of beliefs these are (Sect. 3.2).

1.1.2 Science as Scientia

So much for science in the broadest sense of the term, as 'knowledge'. There is, however, a second use of the word, in which 'science' refers to a more particular kind of knowledge. This sense of the world corresponds to that of the medieval Latin *scientia*, which was itself used to translate the Greek *epistēmē*. Both *scientia* and *epistēmē* referred not just to any body of knowledge, but to a systematic body of beliefs, grounded in reasoning that (ideally) yields certainty, and which show why things are as they are (Pasnau 2010: 23). Medieval thinkers realized that certainty could not always be obtained (Pasnau 2013: 994): in practice the natural philosopher could be content with a lesser degree of confidence (Pasnau 2010: 37). But a body of knowledge was thought of as a *scientia* only if it took the form of a systematized set of general principles which had an explanatory role.

The terms 'systematized' and 'systematic' themselves have a variety of meanings (Hoyningen-Huene 2013: 26–27). But as a first approximation we can say that a body of knowledge is systematic if it 'embodies some kind of order': it is not 'purely random or accidental, it is not chaotic, not arbitrary' (Hoyningen-Huene 2013: 26). A systematic body of knowledge will normally be founded on a set of general principles. It may speak, for example, of the varieties of soil that yield good crops, rather than merely asserting that 'this soil here yields good crops'.

Science as *scientia* will also include 'second-order' reflection, directed not to its target domain (such as animals or plants or heavenly bodies), but to the claims made about it (Elkana 1986: 40). It will, in other words, not merely make claims; it will reflect on what is required for its claims to be true. Second-order reflection is presumably practised in all societies, with regard to at least some beliefs (Elkana 1986: 41). What is characteristic of a science is that such reflection is undertaken in a self-conscious fashion, by means of procedures for testing knowledge claims and generating new knowledge. Those procedures will include principles (such as those of formal logic) against which knowledge claims can be assessed (Sect. 3.4.1).

This use of the word 'science', referring to any systematic body of knowledge, can still be found in modern English. A recent historical work, for example describes the medieval 'liberal arts'—those of the *trivium* (grammar, logic, and rhetoric) and *quadrivium* (arithmetic, geometry, music, and astronomy)—as 'the liberal sciences' (Peltonen 2002: 97). The same idea can be found in German in the use of the term *Wissenschaft*. In the words of a standard dictionary, a *Wissenschaft* is '[an] ordered, logically structured, and coherent field of knowledge'. The academic study of religion, for example, is described in German as *Religionswissenschaft*, while what we call 'linguistics' is *Sprachwissenschaft*. Do all societies have 'science' (or 'sciences') in this sense: systematic bodies of knowledge? That is a question to which I shall briefly return (Sect. 1.3).

1.1.3 Modern Science

We come, finally, to the narrowest sense of the word 'science'. I shall devote an entire chapter to describing the practices to which this use of the term refers. But my initial definition will be a historical one. 'Science' in the narrowest sense of the word refers to the family of knowledge-related practices—what I shall call 'epistemic practices'—that began to take on their present form in seventeenth-century Europe. This definition is modelled on a common understanding of the biological term 'species', in which the criterion for regarding organisms as members of a single species is that they have a common ancestry (Baum and Donoghue 1995: 560). In a similar way, epistemic practices fall into the category of modern science when they, too, have a common ancestry: when they constitute research traditions dating back to developments that occurred in seventeenth-century Europe. Those developments were particularly focused on a study of the physical world by way of measurement, mathematical modelling, and experiment.

Modern science would be regarded by medieval thinkers as just one form of *scientia*, one variety of systematized knowledge. The nearest medieval equivalent was *scientia naturalis*—systematized knowledge of the natural world (Maurer 1985: xv)—which medieval thinkers distinguished from other forms of *scientia*. Thomas Aquinas (ca. 1225–1274), for instance, also regarded theology as a science (*ST* I 1.2), but one that drew its principles from divine revelation rather than from observation and reflection. Today's everyday English usage of the word 'science' resembles the medieval use of *scientia naturalis* in so far as it excludes other forms of systematic inquiry. As W. G. Ward wrote in 1867, 'we shall, for convenience' sake, use the word "science" in the sense which Englishmen so commonly give to it; as expressing physical and experimental science, to the exclusion of theological and metaphysical' (Ward 1867: 255n).

1.2 The Term 'Western'

The next term that requires disambiguation is 'Western'. What meanings does this term have and which of these are intended by those who speak of 'Western science'?

For better or for worse, the idea of 'the West' comes burdened with more than 2000 years of myth, metaphor, and allegory. Within this history, the West has designated a location, a direction, and an idea (Baritz 1961: 619). As the location to which the sun daily descended, the West was naturally associated with death, an idea already found among the Egyptians (Baritz 1961: 620–621). But death could also signify a rebirth to a new life and the West could therefore be a place of promise, a paradise on earth. Such were the Elysian Fields, 'at the ends of the earth', to which—in a tradition established by Homer (*Odyssey* 4.561–568)—heroes could be transported while still alive (Lovejoy and Boas 1965: 291). The biblical tradition of the Garden of Eden 'in the East' (Gen. 2:8) confounded this classical idea, although the two directions could be reconciled by a moment's reflection on the spherical shape of the earth (Baritz 1961: 624). But the West was also the direction of the progress of empire, as in Virgil's account of Aeneas's journey from Troy to Italy to Rome, a journey extended by Geoffrey of Monmouth (ca. 1095–1155) to England (Baritz 1961: 621, 624). Early modern Europeans came to think of the West as the seat of 'God's word and religion', which having come from the East would, with the European voyages of exploration, move eastwards again, to enlighten the New World (Baritz 1961: 635). The idea that Christian enlightenment would move from west to east was picked up by the theologian Jonathan Edwards (1703–1758), but now from a new location, in which the American colonies were 'the West' (Baritz 1961: 637).

It is difficult to know whether these historical and mythological usages continue to influence today's debates. If they do, it may help account for their vigour. But whatever the influence of such mythic themes, it is to today's debates that I must now turn. How are the terms 'the West' and 'Western' used today?

1.2.1 Geographical Use

A first use of the term 'Western' is *geographical*. In this usage, 'Western' refers (in the words of the *Oxford English Dictionary*) to 'any part of the world to the west of one's own region; applied specifically to the Americas and the Caribbean, or to Europe, as opposed to the Middle East and Orient'. As the same dictionary notes, this geographical usage is now merely historical, except when coupled with what I shall call the *cultural* use of the term. It is, in any case, fatally imprecise. As we have just seen, any place is west of any other on a spherical globe. As a geographical term, its only (more or less) precise use is as an indexical, a term (like 'yesterday') whose referent varies from one context to another.

Using the word as an indexical, one can specify its geographical meaning by reference to the European continent, as in the phrase 'Western Europe'. But the boundaries of the region designated in this way are also unclear. The phrase 'Western Europe' became widely used only at the time of the Cold War, when 'an "iron curtain" divided the continent into West and East' (Berger 2017: 16). Once that curtain was lifted, the boundaries of Western Europe were up for negotiation. This was nowhere clearer than in the expansion of the European Union (EU). While membership in the EU had originally been restricted to countries to the West of the 'iron curtain', the changed situation after 1989 opened up this barrier, to allow former 'Eastern European' countries (such as Poland) to gain membership.

1.2.2 Cultural Use

In practice, however, the geographical use of the term 'Western' has rarely been separate from its *cultural* use. In the cultural use of the term 'the West … is not to be found by recourse to a compass' (Berman 1983: 2). It refers to a set of ideas, practices, and institutions whose origins are believed to lie in Western Europe. It is this meaning of the term 'Western' that seems to underlie the use of the phrase 'Western civilization'. In practice, the cultural use of the term 'Western' is most commonly joined with a third, the 'genus and species' use, to which I shall return in a moment.

In the meantime, it will be useful to look more closely at what the cultural use entails.

The idea of the West as a cultural unity—a 'repertoire' of ideas and practices—is relatively new (Bonnett 2004: 25). It emerged both in 'the West' itself—in Europe and North America—and in 'non-Western' societies, particularly those of Asia. While the sense of the idea differed according to the context of its use, the cultural use of the term 'Western' has been rarely (if ever) merely descriptive. It was, and continues to be, heavily laden with judgements of value, originally positive but more recently negative.

In Europe and North America the idea of 'Western' culture emerged (in part) as a response to the decline of the category of 'whiteness'. Nineteenth century writers commonly spoke of a 'white civilization' (Bonnett 2004: 14) to identify what they regarded as a distinctive set of ideas, practices, and institutions. But this racial category was undermined by both class conflict, which divided European societies, and warfare, which set European peoples in conflict with one another. Rather than forming a single civilization, 'white' people seemed divided by political ideology and struggles for power. The more abstract idea of 'the West' offered a way of transcending these differences. Although critics recognized that 'Westerners' often failed to live up to their own ideals, it was the very possession of these ideals that set them apart from other peoples (Bonnett 2004: 28).

While the cultural use of the term 'Western' has always denoted a set of social practices, the practices to which it refers have been many and varied. Nineteenth-century Russia, for example, experienced arguments between 'Westernizers' (*západniki*), who urged the adoption of European ways, and 'Slavophiles', who warned against the rationalism and materialism of 'the West' (Galaktionov and Nikandrov 1967: 25, 28). Consistently with this usage, the Bolshevik revolutionaries initially categorized socialism as 'Western' (Bonnett 2004: 47). But in the aftermath of the Russian revolution and two world wars, there was a narrowing in the range of institutions regarded as Western. In countries outside of the Soviet Union, the West was now thought to be characterized by a capitalist economy and liberal democracy: its ideals were those of 'non-authoritarian government and an open society' (Bonnett 2004: 42). In this context, the United States came to be regarded as the paradigmatically Western country, the leader of an 'Atlantic alliance' (Bavaj 2011: 17), opposing communism, fascism, and

Nazism (Bonnett 2004: 42–43). We hear echoes of this view of the West in a recent *New York Times* opinion piece:

> What we call 'the West' is a centuries-long conversation – Socrates searching for truth, Rembrandt embodying compassion, Locke developing enlightenment liberalism, Francis Bacon pioneering the scientific method. This is our heritage. For all of our history America understood itself as the culmination of the great Western project. (Brooks 2025)

Within the Soviet Union, on the other hand, 'the West' came to be thought of as the home of bourgeois and anti-revolutionary forces, a society characterized by 'decadence', 'imperialism', 'reaction', 'aggression' and 'greed' (Bonnett 2004: 54).

Such was the emergence of the cultural idea of 'the West' in Europe and North America. But the present use of the term owes much of its meaning to thinkers in Asian countries (Bonnett 2004: 81). Here, too, the use of 'Western' was heavily value-laden. At times, it, too, involved a positive evaluation of the practices to which it referred. We find this, for example, among the modernizing Chinese thinkers of the late nineteenth and early twentieth centuries who regarded their own traditional practices as 'backward' (Elman 2003: 104). But Asian thinkers also developed a less positive view of the practices deemed 'Western'. In India, for example, Rabindranath Tagore (1861–1941) argued that Western modernity 'was a misguided form of modernity' (Bonnett 2004: 80), marked by a lack of true spirituality. In a similar way, the Japanese art historian Okakura Kakuzō (1862–1913), often known as Okakura Tenshin, contrasted the Asian love for 'the Ultimate and Universal' with the Westerner's love of the particular and his neglect of the purpose of life (Bonnett 2004: 83). In China, too, the modernizers were opposed by those who were cautious about 'Western' science, arguing that it needed to be complemented by a recovery of China's own moral and spiritual traditions (Elman 2003: 101).

1.2.3 Genus and Species Use

I have spoken of the *geographical* and *cultural* uses of the term 'Western'. But there also exists what I shall call a *genus and species* use. This overlaps the cultural use, since it assumes that the practices called 'Western' have distinctive characteristics. But it also assumes that the practices described as Western are also found, in differing forms, among 'non-Western' peoples. More precisely, it assumes that there are some general categories ('genera') of cultural forms which have both Western and non-Western manifestations (their various 'species'). One can argue, for instance, that all complex societies—those whose activities have become distributed among distinct institutional spheres (Eisenstadt 1964: 376)—have systems of law. On this view, Western legal systems would be a species of a genus, a particular version of an institution also found in other societies.

I mentioned a moment ago the phrase 'Western civilization'. In the days when 'civilization' was used in the singular, to refer to a state of being civilized (as opposed to being a barbarian), there would have been no need to attach the qualifier 'Western'. Civilization was thought to be 'confined to a few privileged peoples or groups, humanity's "elite"' (Braudel 1994: 7), a group to which Europeans were assumed unquestionably to belong. But beginning in the early nineteenth century, the word came to be used in the plural, with talk of differing 'civilizations' (Braudel 1994: 6). In this new usage, the term resembles the anthropologist's use of 'culture'—the idea of different 'civilizations' is akin to that of different 'cultures'—but with reference to complex societies and a focus on historical persistence. When the term 'civilization' is used in this way, the qualifier 'Western' is required, to indicate which species of the genus is being referred to: that which began in Europe or the civilizations of Africa, the Muslim world, or Asia.

1.3 'Western Science'

Drawing together these possible meanings of the terms, what could it mean to call science 'Western'? A first interpretation takes the term 'Western' in its *geographical* sense. Here 'Western' could refer to the region in which science is practised, in the same way as one could speak of 'Chinese agriculture', to refer to the farming practices found in China. But this can hardly be what is intended by the phrase 'Western science'. Science is geographically Western insofar as it took on its present form in a particular region (an idea to which I shall return shortly). But its practice is no longer restricted to any one part of the world. Indeed one of the striking features of modern science has been its ability to be take root and flourish in societies that differ widely with regard to language, customs, and legal and political institutions.

Authors vary in their explanations of this fact. Joseph Needham, for instance, argued that insofar as it relies on observation and mathematical modelling, modern science is inherently 'ecumenical', 'the common property of all men everywhere' (Needham 1978: 112). But this positive view has recently fallen out of favour. It has become more common to claim that science travelled in an alliance with 'capital and political power' as an instrument of Western hegemony (Brockway 1979: 461). But however one explains the geographical spread of modern science, there is no doubt that this has occurred. The practice of modern science is no longer restricted to one part of the globe.

Another way of interpreting the phrase 'Western science' understands 'Western' in a geographical sense, but in the purely historical fashion to which I alluded a moment. On this view, even if science is now practised everywhere, it is Western in origin. While this seems a more plausible claim, it suffers from the ambiguities of the geographical usage. Does 'Western' here mean 'Western European'? The astronomer Tycho Brahe (1546–1601) came from Denmark, while Nicolas Copernicus (1473–1543) came from Poland. Are Denmark and Poland 'Western' countries? Perhaps they are, if we understand the term broadly. But there is more serious objection to this idea, which I shall develop later (Sect. 3.1). It is that modern science was not a *creatio e nihilo*. The pioneers of modern science wove together existing ideas and practices into new institutional forms and the ideas and practices they wove together came from far beyond whatever boundaries one can plausibly assign to 'the West'.

A second interpretation of the phrase 'Western science' understands 'Western' in a *cultural* sense. On this view, one can admit that the origins of modern science are not purely Western, while still insisting that it has features that are not shared by epistemic practices in other societies. It is this cultural use of 'Western' that appears to be the most common one: those who speak of 'Western science' generally do so in order to contrast it with other forms of knowledge. One author, for example, claims that the Chinese view of causality is very different from that found in 'Western science' (Cheng 1976: 4), the former being 'holistic' while the latter is 'atomistic' and 'mechanistic' (Cheng 1976: 18). Other writers argue that because of its particular character, the scope of Western science is limited. While it may be 'useful for prediction, control, and the design of manipulative technologies', it 'is in no way qualified to provide a worldview adequate to guide individual and societal decisions' (Harman 1996: 31).

Is science Western by origin? Is science Western in character? Those are the two questions to which this study will be devoted, the first question being addressed in Chap. 3 and the second in Chap. 4. The first is a historical question, asking about the origin and development of modern science in relation to European societies. The second is an analytical question, asking about the character of the practices we customarily call 'scientific'. The answers I give to these two questions will be interwoven, insofar as the history has shaped the character. But I shall treat them separately for the sake of clarity.

What about the third sense of 'Western', the *genus and species* use of the term? Any argument that science is Western in origin or character assumes that it differs in some respects from epistemic practices found in non-Western societies. But today's postcolonial writers often make a further claim, arguing that other societies have their own form of science. Insofar as these are thought of as differing forms of science, this *genus and species* usage overlaps the *cultural* use. Insofar as it distinguishes the sciences that are alleged to have developed in differing parts of the world, this usage overlaps the *geographical* use. What is distinctive about the *genus and species* usage is its assumption that there are epistemic practices comparable to modern science elsewhere.

Such practices could, of course, be simply variant forms of the science which began to take on its current form in early modern Europe. This appears to have been what Stalinist thinkers meant when they contrasted 'Western science' with 'Soviet science' (Krementsov 1997: 179). They were suggesting that the modern scientific tradition had taken on different forms under capitalism and communism. But the more interesting question is whether there exist epistemic practices comparable to modern science that have developed independently from the science found in 'the West'.

It is crucial, in answering this question, to keep in mind the three senses of the term 'science' (Sect. 1.1). A society could have well-founded bodies of knowledge without having science in the second and narrower sense of the word: that of the Latin *scientia* (or the German *Wissenschaft*). It has been argued, for example, that the development of systematic bodies of knowledge is practically inconceivable in societies that lack writing (Goody 1977: 43; Lloyd 1990: 132–133). But even if all societies have something akin to *scientia*, they may not have developed the particular form of *scientia* I am calling 'modern science'.

There have been clear instances of 'sciences' in the sense of the Latin *scientia* (or the German *Wissenschaft*) in at least some non-Western societies. An early example would be the astronomical (and astrological) sciences developed in ancient Mesopotamia (Rochberg 2018: 9), to which elements of modern 'Western' science can be traced (Sect. 3.1.1). A second example is the correlative cosmology developed in ancient China, which continues to serve as a foundation for Chinese traditional medicine (Bodde 1991: 11–12). A third example is the collection of disciplines that developed in India known as *śāstras* (Pollock 1989: 402). Insofar as these are systematically arranged bodies of knowledge that can have an explanatory role, they count as instances of what medieval writers called *scientia* (Pingree 1992: 559).

What about science in the third and narrowest sense of the term? Do other societies have epistemic practices comparable to modern science? What I shall be arguing in the chapters that follow is that the elements of modern scientific practice are found in many non-Western societies. As it happens, they came together into the particular institutional forms that English speakers call 'science' in Western Europe. Did a comparable coming together happen in any other cultural context? At first sight, it would appear that it did not, that these practices came together only in Western Europe as a result of peculiarly European social and political developments (Sect. 4.2). But if (as I shall argue) the practices of modern science are refined versions of epistemic practices found in all societies (Sect. 3.3), this will help bridge any apparent gap between 'the West' and 'the Rest'.

References

Baritz, Loren. 1961. The Idea of the West. *The American Historical Review* 66 (3): 618–640.

Barnhardt, Ray, and Angayuqaq Oscar Kawagley. 2005. Indigenous Knowledge Systems and Alaska Native Ways of Knowing. *Anthropology and Education Quarterly* 36 (1): 8–23.

Baum, David A., and Michael J. Donoghue. 1995. Choosing Among Alternative "Phylogenetic" Species Concepts. *Systematic Botany* 20 (4): 560–573.

Bavaj, Riccardo. 2011. "The West": A Conceptual Exploration. *Europäische Geschichte Online* http://www.ieg-ego.eu/bavajr-2011-en.

Beiser, Frederick C. 1992. *Revolution and Romanticism: The Genesis of Modern German Political Thought, 1790–1800.* Cambridge, MA: Harvard University Press.

Beiser, Frederick C. 2011. *The German Historicist Tradition.* New York: Oxford University Press.

Berger, Stefan. 2017. Western Europe. In *European Regions and Boundaries: A Conceptual History*, ed. Diana Mishkova and Balázs Trencsényi, 15–35. New York: Berghahn.

Berlin, Isaiah. 2013. In *Three Critics of the Enlightenment: Vico, Hamann, Herder*, ed. Henry Hardy, 2nd ed. Princeton, NJ: Princeton University Press.

Berman, Harold J. 1983. *Law and Revolution: The Formation of the Western Legal Tradition*. Cambridge, MA: Harvard University Press.

Bodde, Derk. 1991. *Chinese Thought, Society, and Science: The Intellectual and Social Background of Science and Technology in Pre-modern China*. Honolulu, HI: University of Hawaii Press.

Bonnett, Alastair. 2004. *The Idea of the West: Culture, Politics and History*. London: Palgrave Macmillan.

Braudel, Fernand. 1994. *A History of Civilizations* (1987). Translated by Richard Mayne. London: Penguin Books.

Brockway, Lucile H. 1979. Science and Colonial Expansion: The Role of the British Royal Botanic Gardens. *American Ethnologist* 6 (3): 449–465.

Brooks, David. 2025. America is Surrounded by Enemies … That We Created. *The New York Times* 13 March (Section A): 25.

Canagarajah, Suresh. 2002. Reconstructing Local Knowledge. *Journal of Language, Identity, and Education* 1 (4): 243–259.

Cheng, Chung-ying. 1976. Model of Causality in Chinese Philosophy: A Comparative Study. *Philosophy East and West* 26 (1): 3–20.

Durie, Mason. 2004. Understanding Health and Illness: Research at the Interface Between Science and Indigenous Knowledge. *International Journal of Epidemiology* 33 (5): 1138–1143.

Eisenstadt, Shmuel N. 1964. Social Change, Differentiation and Evolution. *American Sociological Review* 29 (3): 375–386.

Elkana, Yehuda. 1986. The Emergence of Second-Order Thinking in Classical Greece. In *The Origins and Diversity of Axial Age Civilizations*, ed. S. N. Eisenstadt, 40–64. Albany, NY: State University of New York Press.

Elman, Benjamin A. 2003. "Universal Science" Versus "Chinese Science": The Changing Identity of Natural Studies in China, 1850–1930. *Historiography East and West* 1 (1): 68–116.

Galaktionov, A. A., and P. F. Nikandrov. 1967. Slavophilism, its National Roots and its Place in the History of Russian Thought. *Soviet Studies in Philosophy* 6 (2): 22–32.

Goody, Jack. 1977. *The Domestication of the Savage Mind*. Cambridge: Cambridge University Press.

Greene, John C. 1981. *Science, Ideology, and World View: Essays in the History of Evolutionary Ideas*. Berkeley, CA: University of California Press.

Harman, Willis W. 1996. The Shortcomings of Western Science. *Qualitative Inquiry* 2 (1): 30–38.

Hoyningen-Huene, Paul. 2013. *Systematicity: The Nature of Science. Oxford Studies in Philosophy of Science*. Oxford: Oxford University Press.

Krementsov, Nikolai. 1997. *Soviet Science*. Princeton, NJ: Princeton University Press.
Landgrebe, Ludwig. 1940. The World as a Phenomenological Problem. *Philosophy and Phenomenological Research* 1 (1): 38–58.
Leerssen, Joep. 2004. Literary Historicism: Romanticism, Philologists, and the Presence of the Past. *MLQ: Modern Language Quarterly* 65 (2): 221–243.
Lloyd, G. E. R. 1990. *Demystifying Mentalities*. In *Themes in the Social Sciences*. Cambridge: Cambridge University Press.
Lovejoy, Arthur O., and George Boas. 1965. *Primitivism and Related Ideas in Antiquity*. In *Contributions to the History of Primitivism*. New York: Octagon Books.
Maurer, Armand. 1985. Introduction. In *St Thomas Aquinas, The Division and Methods of the Sciences – Questions V and VI of his Commentary on the De Trinitate of Boethius*, ed. Armand Maurer, 4th ed., vii–xli. Toronto: Pontifical Institute of Medieval Studies.
Mignolo, Walter. 2012. Decolonizing Western Epistemology/Building Decolonial Epistemologies. In *Decolonizing Epistemologies*, ed. Ada María Isasi-Díaz and Eduardo Mendieta, 19–43. New York: Fordham University Press.
Needham, Joseph. 1978. Address to the Opening Session of the XV International Congress of the History of Science, Edinburgh, 11 August 1977. *The British Journal for the History of Science* 11 (2): 103–113.
Nozick, Robert. 1981. *Philosophical Explanations*. Cambridge, MA: Harvard University Press.
Okere, Theophilus. 2005. Is There One Science, Western Science? *Africa Development* 30 (3): 20–34.
Pasnau, Robert. 2010. Medieval Social Epistemology: *Scientia* for Mere Mortals. *Episteme* 7 (1): 23–41.
Pasnau, Robert. 2013. Epistemology Idealized. *Mind* 122 (488): 987–1021.
Pelto, Pertti J., and Gretel H. Pelto. 1997. Studying Knowledge, Culture, and Behavior in Applied Medical Anthropology. *Medical Anthropology Quarterly NS* 11 (2): 147–163.
Peltonen, Markku. 2002. Citizenship and Republicanism in Elizabethan England. In *Republicanism: Volume 1, Republicanism and Constitutionalism in Early Modern Europe: A Shared European Heritage*, ed. Martin van Gelderen and Quentin Skinner, vol. 1, 85–106. Cambridge: Cambridge University Press.
Pingree, David. 1992. Hellenophilia versus the History of Science. *Isis* 83 (4): 554–563.
Pollock, Sheldon. 1989. The Idea of Śastra in Traditional India. In *Shastric Traditions in Indian Arts*, ed. Anna Libera Dallapiccola, 17–26. Stuttgart: Steiner Verlag.

Rochberg, Francesca. 2018. Science and Ancient Mesopotamia. In *The Cambridge Companion to Science*, ed. Alexander Jones and Liba Taub, vol. 1, 7–28. Cambridge: Cambridge University Press.

Ryn, Claes G. 2003. *A Common Human Ground: Universality and Particularity in a Multicultural World*. Columbia, MO: University of Missouri Press.

Seuffert, Nan. 1997. Circumscribing Knowledge in Aotearoa/New Zealand: Just Epistemology. *Yearbook of New Zealand Jurisprudence 1*:97–125.

Sikka, Sonia. 2005. Enlightened Relativism: The Case of Herder. *Philosophy and Social Criticism* 31 (3): 309–341.

Ward, W. G. 1867. Science, Prayer, Free will, and Miracles. In *The Dublin Review*, April ed., 255–298.

Waters, Lindsay. 2001. The Age of Incommensurability. *Boundary 2* 28 (2): 133–172.

Watson-Verran, Helen, and David Turnbull. 1995. Science and Other Indigenous Knowledge Systems. In *Handbook of Science and Technology Studies*, ed. Sheila Jasanoff et al., revised ed., 115–139. London: Sage Publications.

CHAPTER 2

What Do We Mean by 'Science'?

Abstract What are the characteristics of the modern sciences? Efforts to define 'science'—to find necessary and sufficient conditions for the use of the word—are generally thought to have failed. But the word can still be used as a 'family-resemblance' term. Even if there is no characteristic shared by all the sciences, there are features found among them by which they resemble one another. Within each science, we can distinguish between individual and social practices, the former being those that can be engaged in (at least in principle) by individuals. We can also distinguish four levels of scientific practice: the operational, the phenomenal, the theoretical, and the metatheoretical. The question 'Is science Western?' has different answers, depending on which of these levels one focuses on.

Keywords Family-resemblance • Paradigms • Pseudoscience • *Themata* • Theory-ladenness

In my introductory chapter, I tried to narrow the various uses of the word 'science' so as to arrive at the common English use of the term. In this usage, the word refers to what I am calling 'modern science', which I defined as the family of epistemic practices that began to take on their present form in seventeenth-century Europe (Sect. 1.1). But while this historical definition is a useful first approximation, it does not tell us very much. Can I be more specific about what this set of practices entails?

G. W. Dawes, *Is Science Western?*,
https://doi.org/10.1007/978-3-032-10641-4_2

2.1 A 'Family-Resemblance' Term

The problem here is that it is notoriously difficult to identify what characteristics make a practice a science. Twentieth-century philosophers devoted a lot of energy to finding a 'demarcation criterion': a feature that would single out a practice as a science, the lack of which would relegate a practice to the category of 'non-science', or (if it claimed to be a science) 'pseudoscience'. I am sympathetic to this project, for there are good reasons to want to make such distinctions. There are practices that claim the mantle of science—'piggy-backing', as it were, on science's reputation for reliability—whose credentials for doing so are dubious. Some young-earth creationists, for example, who believe the world was created by God less than 15,000 years ago, defend what they call 'creation science', arguing that it should be taught as an alternative to mainstream science in public schools. But many of us would want to argue that creation science is not a science, at least not in the same sense as physics or genetics.

Let me begin with the idea of a pseudoscience (of which 'creation science' is arguably an example). The problem with a pseudoscience is that it claims to be a science while lacking the degree of reliability that sciences can attain. So perhaps we could reformulate our quest to define a science, describing it as a quest to distinguish between reliable and unreliable ways of attaining knowledge. But this will not work, for such a distinction does not correspond to that between the scientific and the non-scientific. There are many ways of attaining knowledge no one would call 'scientific' that we still regard as reliable. If I glance out my window to see if it is still raining (as I just have), I can gain knowledge, but no one would think of this as a scientific experiment. Whether science also relies on such everyday ways of attaining knowledge is a question to which I shall return (Sect. 3.3). But the point to be made here is that we cannot simply identify non-scientific practices with unreliable ones.

Even if we regard at least some everyday practices as reliable, scientific practices arguably have a higher degree of reliability (Sect. 3.4). So what makes them particularly reliable? After the abandonment of the ancient idea that science gives rise to certain knowledge rather than mere opinion, attempts to make a distinction between science and non-science focused on method. But philosophers could not agree on what this alleged 'scientific method' is (Laudan 1983: 112–116). One famous example was the argument of the 'logical empiricists' associated with the Vienna Circle of the 1920s and 1930s. These thinkers claimed that for a statement to be scientific, it had to be empirically verifiable. The problem here is that no purportedly universal law is verifiable (*as* a universal law), for there may always be an (as yet unobserved) exception. Nor will Karl Popper's criterion of falsifiability serve us to distinguish science from non-science. Much everyday knowledge is falsifiable: I could believe it is still raining and look out the window to discover it is not. More seriously, even many apparently pseudoscientific claims (such as those made by creation scientists) are falsifiable, in the sense that one can conceive of observations that would prove them false (Laudan 1983: 120–123).

Such were the arguments put forward by Larry Laudan in 1983. There is, however, another difficulty that faces this project, which arises from the fact that the sciences employ refined versions of epistemic practices we employ in everyday life (Sect. 3.4). (Looking through a telescope has *something* in common with the practice of looking out a window.) If this is true, a sharp distinction between science and non-science seems impossible. When it comes to reliable-making features, everyday and scientific practices may lie on a continuum. But even if we give up the quest to find necessary and sufficient conditions for regarding a practice as a science, it does not mean we cannot think of the sciences as forming a category, albeit one with no clear boundaries. The different sciences can resemble one another as do the members of a human family, where a son may resemble his father in some respects, and his sister in other respects, and so on (Pigliucci 2013: 21–22).

When it comes to modern science, a list of such 'family resemblance' properties might include the following. The modern sciences have each achieved some degree of systematicity, along a range of dimensions (Hoyningen-Huene 2013: 29). They make claims to knowledge that take the form of general principles. These principles (or at last many of them) are linked to one another by inferential relations in ways that allow them to be checked for consistency and tested against experience. Many sciences employ mathematical modelling and test their claims by way of controlled experiments. Insofar as they do share such features, the sciences are also dependent on one another in various ways (Hoyningen-Huene 2013: 120). Archaeology may use methods of dating artifacts that are dependent on physics. Sociology may use statistical methods that resemble those employed in population biology. These features may not offer a clear way of distinguishing science from non-science. But they do give some substantive meaning to the phrase 'modern science'.

2.2 Social and Individual Practice

So much for what I mean by 'modern science'. But it will also be useful to make some distinctions regarding the kinds of practices that make up a science. A first such distinction is related to one made by Nikolai Krementsov in his study of Soviet science. Krementsov distinguished two kinds of scientific practice, which he called the 'social' and the 'experimental'. The experimental he described as all those practices that produce 'scientific knowledge, facts and artifacts, hypotheses and theories' (Krementsov 1997: 307 n. 4). The social, by way of contrast, consists of those practices relating to 'formal and informal organizations, patrons, careers, roles, and policies' (Krementsov 1997: 307 n. 4). The first distinction I want to make in this context resembles Krementsov's, being a distinction between *social* and *individual* scientific practices. There are practices that scientists can engage in only collectively and those they can engage in when acting as individuals.

In making this distinction I want to distance myself from a long-standing philosophical prejudice, which Krementsov seems to share. It is the assumption that it is the individual practices scientists engage in that do the epistemic work (as it were), producing knowledge. The social practices of scientific communities, on this view, are thought to be either neutral or a source of error. We find this idea in Robert Nola and Howard Sankey's discussion of what they call the 'personal, professional and political interests' of individual scientists. These interests are, they argue, 'randomly linked to whether the theories chosen on the basis of such interests ... are successful' (Nola and Sankey 2000: 7). But this seems wrong. The personal and professional interests of scientists are not randomly associated with the success of science; they can make a positive contribution to it.

An argument to this effect was made as long ago as 1968 by the sociologist Robert Merton. Merton noted that scientific communities have systems of rewards for intellectual achievement, rewards in terms of both money and social status. He argued that while apparently 'extrinsic' to the actual work of scientific research, such rewards encourage work of a high quality as well as further effort on the part of those who receive them (Merton 1973: 440–441). Of course, this reward system has its downside. To those who have will be given more—'eminent scientists get disproportionately great credit for their contributions' (Merton 1973: 443)—while the lure of attaining credit can encourage fraud (Merton 1973: 309;

O'Connor et al. 2023: Sect. 4.2). But it remains the case that even factors one might expect to interfere with scientific progress—such as 'the thirst for fame and fortune'—can actually play a constructive role in the quest to attain knowledge (Kitcher 1990: 16).

That is one aspect of the social dimension of science, which as to do with 'extrinsic' motivation and rewards. But a second aspect consists in the fact that many of the practices that yield scientific knowledge are collective practices. Much scientific research is done by individuals working together to achieve results that none could achieve working alone (Merton 1973: 328). Even the work of individuals is improved by peer review and tested by attempts to replicate their results. Again, these procedures are far from perfect. There exists, for example, a 'replication crisis' in the sciences, consisting in the fact that many results are not subject to experimental testing (Romero 2019: 3). Worse still, when attempts are made to replicate findings that have become widely accepted in a field, those attempts have sometimes failed (Romero 2019: 3). But efforts are being expended to address these problems (Korbmacher et al. 2023), precisely because the trustworthiness of science depends on reliable collective procedures.

It follows that scientific knowledge does not emerge merely from individual achievement; it arises from the collective procedures of a certain kind of community. It is for this reason that the physicist John Ziman argued that it was difficult to conceive of a Robinson Crusoe scientist, for 'scientific research is a social activity' (Ziman 1968: 10). We should not overstate this idea. A scientist marooned alone on an island could engage in individual practices. She observe carefully, could classify what is observed, take measurements, and perform simple experiments. What such a person could not do is discuss her findings with others, send a paper off for peer review, apply for grants that would free her from the need to gather food, or engage in complex experiments that require multiple workers. The former set of practices are what I am calling individual practices. The practices a Robinson Crusoe scientist could not perform are the social ones.

2.3 Dimensions of Scientific Practice

There is, however, a second set of distinctions that will be useful in this context. I have already argued that 'science' is an family-resemblance term, covering a range of different disciplines—as diverse as physics, sociology, and zoology—that employ related but not identical assumptions and methods of inquiry. In this sense modern science is clearly not 'one thing'. But none of these individual sciences is itself 'one thing': the practice of each is multidimensional. No science can be understood by reference to just one of its dimensions alone.

2.3.1 *Levels of Scientific Practice*

Philosophers often write as though scientists confront the world directly, drawing from unmediated acts of observation the complex theories they employ to explain it. But the process by which the sciences make contact with their objects of study is a complex one. Scientists articulate their theories only after engaging in distinctive forms of activity, such as sampling, classifying, analysing, and experimenting. These activities form a series of linked mediations bridging the apparent gap between observer and world.

Bruno Latour illustrates this feature of science by way of a study of a soil science field trip in the Amazon. Fieldwork in the wild, Latour argues, would be impossible for an observer employing nothing more than what Kant would have called '*a priori* forms of intuition' (Latour 1999: 44). Plants and soil must be turned into samples by being placed into boxes or bags and numbered, the numbers being traceable to the original context by a notebook. Those samples are then taken back to the laboratory and transformed—plant samples, for instance, are dried to remove fungi—before being further analysed. The process of analysis itself involves a movement from the physical geometry of the sampling boxes, in which individual samples can be compared and arranged, to the more abstract geometry of the chart or diagram on paper. At each step the object of analysis is transformed and key to these transformations is the use of standardized measures (Sect. 3.4.2). A colour chart, for example, can enables the observed colour of a soil sample to be given a numerical value. As simple tool such as this, Latour writes, enables the researcher 'to cross the threshold between local and global' (Latour 1999: 59), laying the foundation for the establishment of theoretical laws.

In a word, what scientists call 'data' are not simple observations: they are the outcome of processes that reshape what is observed so as to reveal its theoretical significance. The final product of scientific inquiry—a theoretical conclusion—refers to the reality being studied in the sense that we can trace these transformations back to their source (Latour 1999: 69). If we accept this idea, then we can think of this chain of transformations as forming 'levels', leading from the more concrete and particular to the more abstract and theoretical.

One attempt to identify these levels can be found in the work of Hans Radder, who distinguishes between the experimental and theoretical dimensions of scientific practice (Radder 2012: 64), further distinguishing (within the theoretical level) 'conceptual-theoretical work' and 'formal-mathematical activities' (Radder 2012: 111). But a still more useful model is that offered by Srđan Lelas. On Lelas's account, the most immediately engaged level of scientific activity is the *operational*, which is the realm of the practical activities that Latour focuses on: sampling, classifying, experimenting, and so on. A second level is that of *phenomena*: those observable facts that are revealed—or in some cases produced (Hacking 1983: 224–225)—at the level of the operational. A third level is the *theoretical*, in which these phenomena can be explained by reference to 'more general principles, laws, and concepts' (Lelas 2000: 239). A final level is that of the *meta-theoretical*: the realm of the 'framework theories' that define the nature of the inquiry and the kinds of objects it contains (Wellman and Gelman 1992: 342).

A few comments on the last of these levels—the metatheoretical—may be in order, since it will feature prominently in the discussion that follows. One of the goals of Thomas Kuhn's talk of 'paradigms' was to draw attention to the fact that scientific theories rest on metatheoretical assumptions. A 'paradigm shift' is (among other things) a change of metatheoretical assumptions. Aristotle's physics, for instance, relied on assumptions about 'natural' motion that Galileo abandoned (Kuhn 1970: 119). But Kuhn's category of a 'paradigm' is notoriously loose, and mixes elements from all four of the levels Lelas distinguishes (Kuhn 1970: 10). So a more useful way of thinking about the metatheoretical level of science is to think of it as the domain of what Gerald Holton calls *themata*. *Themata*, Holton writes, are 'basic … presuppositions' that are 'not directly derivable from the phenomena and are not [directly] provable or falsifiable' (Holton 1996: 455). But although not directly testable against scientific data, *themata* function, whether consciously or unconsciously, to shape a scientist's thought. *Themata* are presuppositions in the sense that they 'are usually not explicitly acknowledged, they are not explicitly taught, and they are not listed in the research journals or textbooks' (Holton 1996: 455). If they are articulated, as they sometimes are (Sect. 3.4.1), it will normally be during a period of crisis, when an established mode of theorizing is under threat.

Many themata are best regarded as methodological: they specify how the the practice of a science ought to proceed. *Themata* of this kind include the principles of uniformity and analogy in the historical sciences, the first articulated by Charles Lyell with regard to geology (Lyell 1835: 331), the second by Ernst Troeltsch with regard to history (Troeltsch 1991: 13–14). Both can be regarded as expressions in particular sciences of a more general *thema*, namely the principle of the *vera causa* (true cause): the idea that a posited cause must belong to a category of causes of which we have independent evidence (Mill 1882: 612 [3.14.4]). Themata are not, however, dogmas that are immune to criticism. A period of crisis, when a science appears to be making no progress, may lead to its *themata* being called into question. Einstein's adherence to the traditional *thema* he called 'strict causality' (Ben-Menahem 1993: 299) led him to reject the 'inherent probabilism' of Bohr's quantum mechanics (Holton 1996: 457). But it was Bohr's revolutionary physics that eventually prevailed.

2.3.2 Practice as Theory-Laden

Are these distinctions between levels of scientific practice tenable? In particular, can we distinguish between the operational and the theoretical level of scientific practice? Is it not the case that even the most fundamental level of scientific practice—that of observation—is 'theory-laden' (Hanson 1958: 19)? Well, yes, it is. But we need to think carefully about what this does, and does not, entail.

There are clear cases in which a scientific practice rests on a body of theory. When, for instance, a biologist uses an electron microscope to observe the structure of a cell, her practice assumes certain facts about the behaviour of electrons in electromagnetic fields (Carrier 1994: 9). But in most such cases we can distinguish between the theory of the instrument that is being employed and the theory that is being tested (Franklin 2015: 156). Provided that these differ, we can describe the first as belonging to the operational level of scientific practice and the second to the theoretical.

What about 'bare' acts of observation: those that do not employ instruments? The earliest stages of visual perception appear to be driven 'bottom-up' by the data registered by the retina (Raftopoulos 2001: 428). Some early stages of visual processing do employ 'general world principles' (Raftopoulos 2001: 429). But these are uncontroversial. If not strictly innate, they are principles we are born predisposed to employ (Sect. 3.3.3), What we call 'observation', however, goes beyond these early states of vision, since it involves the identification of what is perceived (Raftopoulos 2001: 425). When we report an observation we are applying concepts to the data of perception: we are describing what we perceive in a particular way (Lindsay and Norman 1977: 251). So the question then arises: what kinds of of concepts are involved in observation reports?

Some such concepts involved in observation reports can be regarded as 'theoretical'. Take, for instance, a physicist watching tracks in a cloud chamber. Her theoretical knowledge will enable her to 'see' the decay of subatomic particles, where an untrained observer would see only lines (Franklin 2015: 155). But the assumptions involved in many acts of observing—such as seeing whether an indicator light is 'on' or 'off' or whether a heated substance has changed colour—are not helpfully thought of as 'theoretical' (Heidelberger 2003: 146). They are (once again) those general assumptions about the world we born predisposed to apply. Here, too, we can distinguish between the level of practice and that of theory, since observers holding very different theoretical views will be able to agree on what is being observed.

Much more could be said regarding the relation between theory and practice in science. Take, for instance, cases in which observations rely on the theory they are supposed to be testing. While this may appear to be a viciously circular process, it is not necessarily so, for there is still no guarantee that the observations will confirm the theory (Chang 2004: 95). But these few remarks suggest that I need not abandon my distinction between the operational level of science, which is more or less laden with differing kinds of theory, and the level at which theoretical claims are formulated and discussed. While the levels of scientific practice continually interact, we can distinguish the activities characteristic of each level.

Not only can we distinguish them; we can see that these levels of practice enjoy a degree of autonomy. They enjoy a degree of autonomy, first of all in the sense that a theory can be devised in advance of its experimental testing. Einstein, for instance, appears to have arrived at his theory of special relativity 'primarily on nonempirical grounds' (Gutting 1972: 66), although he eagerly awaited experimental tests (Hentschel 1992: 618–619). But the operational level of science also enjoys a degree of autonomy. This fact has been highlighted by 'the new experimentalists', a group of philosophers who argue that experimentation can have 'a life of its own' (Hacking 1983: 150). Firstly, experiments can be conducted for a range of purposes, only one of which is the testing of theories. An experiment may, for instance, be performed to distinguish a real effect from an artefact (Mayo 1994: 270–271). Secondly, experimental results may be justified independently of theory (Mayo 1994: 271). It may be, for instance, that the only 'theory' on which they rely consists of the uncontroversial assumptions of which I have already spoken. Thirdly, experimental results can persist through changes of theory (Mayo 1994: 271). The experiments performed by David Brewster (1781–1868) with polarized light, for instance, produced results that have survived radical changes in optical theory (Hacking 1983: 157).

In his study of the history of the measurement of temperature, Hasok Chang offers another example of the partial autonomy of the operational level of science. As Chang writes,

> by the middle of the nineteenth century temperature became measurable in a coherent and precise manner over a broad range, but all of that was achieved without much theoretical understanding. It is not that there was a complete lack of relevant theories – there had always been theories about the nature of heat, since ancient times. But until the late nineteenth century no theories of heat were successful in directing the practice of thermometry in a useful way. (Chang 2004: 159)

Measuring temperature, in other words, could be an established scientific practice even before there was an agreed theory of what 'temperature' refers to. The measurement of temperature could become an established scientific practice because ways had been found of establishing its reliability that depended on very few theoretical assumptions (Chang 2004: 94).

2.3.3 *The Primacy of the Operational*

We can go further than distinguishing these levels of scientific practice. We can also can speak of *the primacy of the operational* level. The theoretical and metatheoretical levels of science are gradually reshaped by phenomena that are revealed, or (in some cases) produced on the operational. This may seem an unremarkable idea. It amounts to little more than the claim that scientific theories are ultimately answerable to the data they seek to explain. But it is still worth noting.

The primacy of the operational and phenomenal levels of science has been highlighted in studies by William Brewer and his colleagues. At first sight, their work seems to reinforce the 'theory-ladenness' claim, for it highlights the 'top-down' influence of theory on scientific practice. This influence, they argue, is not restricted to observation, in the sense of visual perception. It also affects (a) attention, (b) data interpretation, (c) data production, (d) memory, and (e) scientific communication (Brewer and Lambert 2001: S177). Scientists are also unlikely to report observations inconsistent with their theoretical expectations and are likely to interpret ambiguous or degraded data in ways consistent with their assumptions (Brewer and Loschky 2005: 38). A famous example is the apparent discovery of 'N-rays' by the French physicist Prosper-René Blondlot in 1903, where theoretical expectations lead Blondlot and others to perceive differences in the intensity of a spark when none existed (Brewer and Loschky 2005: 39–40).

Yet Brewer and his colleagues also argue that this 'top-down' influence is largely limited to cases of ambiguous data (Brewer and Lambert 2001: S184), which (as it happens) the methods of science are designed to avoid (Raftopoulos 2001: 434–435). Where the data are unambiguous, 'bottom-up factors are the overwhelming influence on what we perceive' (Brewer and Loschky 2005: 43). In a similar way, studies of human perception more generally suggest that 'when bottom-up information conflicts with top-down information, the former usually dominates' (Barsalou 1999: 589).

This conclusion may appear to be challenged by a recent development in cognitive science, namely the 'predictive processing' model of perception. According to this model, the human perceptual system does not (normally) operate by gradually piecing together information from the environment to form a picture of its world. Rather, it predicts what it will perceive by using a 'generative model' of its world, employing action and perception as a way of correcting its predictions (Clark 2016: 3).

At first sight, this model may appear to undermine the objectivity of scientific investigation, for it suggests that 'what we (think we) perceive may – courtesy of all that top-down influence – become … infected by what we expect to perceive' (Clark 2016: 199). But this worry is misplaced. Not only are individual predictions corrected by the inputs of the senses, but the generative model we use has been trained—by both natural selection and learning—to produce predictions that are sufficiently accurate for most human needs. Note, too, that while error correction seems to be a central feature of everyday perception, it is particularly important in the sciences (Sect. 3.4.1), which have developed practices designed to counter inappropriate forms of theory-driven influence (Brewer and Loschky 2005: 44). Those who believed they had discovered N-rays may have realized their mistake sooner had they performed controlled experiments (Lagemann 1977: 282), of the kind that had been recommended by John Stuart Mill (1806–1873).

To these theoretical considerations I can add a historical one, for the pioneers of modern science laid great stress upon the primacy of the operational level of scientific practice. Of particular importance was the practice of experimentation. In England, the new science came to be known as 'experimental philosophy' (Anstey 2005: 215), whose advocates insisted that experimentation could be used to demonstrate the existence of laws of nature even though the causes of such phenomena remain unknown (Henry 1986: 359). Universal gravitation was the clearest example. It discoverer, Isaac Newton (1642–1727), insisted that it 'really exists and acts according to the laws that we have set forth' (Newton 1999: 943 [*P* III GS]), even though he was unable to discover its cause (Henry 1986: 358).

Similar remarks can be made about what I shall later describe as 'the mechanical philosophy'. Although early modern science was strongly influenced by the metaphor of the machine (Sect. 4.3.1), this preference did not always involve a commitment to a mechanistic metaphysics. For many of early modern scientists—René Descartes (1596–1650) being a possible exception—the use of the machine metaphor was what Dimitri Levitin calls 'operational' (Levitin 2022: 112). Such thinkers used mechanical principles to explain phenomena, without being committed to the idea that the underlying reality is *merely* mechanical. As Robert Hooke in 1681, speaking about light, 'provided we know what the motion is, and the Rules, Powers, and Proportions of that, we need not much consider the Substance of it' (Levitin 2022: 113 n. 349).

Nor was this attitude restricted to England. In Italy, Galileo had already insisted that the focus of his 'new science' was what could be demonstrated mathematically and verified experimentally. Theories about the underlying reality of the world—what is sometimes called 'matter theory'—he regarded with suspicion. In Galileo's *Dialogues Concerning Two New Sciences*, his spokesperson, Salviati, simply sets aside the question of the cause of the acceleration of falling bodies, arguing that it will suffice 'to investigate and demonstrate some attributes of a motion so accelerated' (Galilei 1989: 158–159). Marin Mersenne (1588–1648), best known for his music theory, echoed Galileo in this respect, saying that since we cannot know the underlying causes of natural phenomena, all we should expect of scientists are 'faithful reports of their observations and experiment' (Garber 2013: 16).

We can, then, distinguish levels of scientific practice—operational, phenomenal, theoretical, and metatheoretical—and among these levels it is the operational and phenomenal that have priority, acting as a check on theorizing and sometimes operating in the absence of theorizing. In the pages that follow I shall argue that each of these levels can stand in a different relation to its social and historical context. This complicates any attempt to say whether modern science is Western. Whether we think of 'Western' as Western in origin or Western in character (Sect. 1.3), one level of scientific practice may have distinctively Western features, while another may involve practices that are found in all human societies.

I shall argue that this is indeed the case. At its operational level modern science involves refined versions of epistemic practices that are very ancient. In their most basic forms, they are practices in which all humans engage. It follows that, at least at its operational level, the set of practices we call modern science is not exclusively Western. If modern science has distinctively Western features, these are more evident on its theoretical and metatheoretical levels.

References

Anstey, Peter R. 2005. Experimental versus Speculative Natural Philosophy. In *The Science of Nature in the Seventeenth Century: Patterns of Change in Early Modern Natural Philosophy*, Studies in History and Philosophy of Science, ed. Peter R. Anstey and John A. Schuster, vol. 19, 215–242. Dordrecht: Springer.

Barsalou, Lawrence W. 1999. Perceptual Symbol Systems. *Behavioral and Brain Sciences* 22 (4): 577–660.

Ben-Menahem, Yemima. 1993. Struggling with Causality: Einstein's Case. *Science in Context* 6 (1): 291–310.

Brewer, William F., and Bruce L. Lambert. 2001. The Theory-Ladenness of Observation and the Theory-Ladenness of the Rest of the Scientific Process. *Philosophy of Science* 68 (3): S176–S186.

Brewer, William F., and Lester Loschky. 2005. Top-down and Bottom-up Influences on Observation: Evidence from Cognitive Psychology and the History of Science. In *Cognitive Penetrability of Perception: Attention, Action, Strategies, and Bottom-up Constraints*, ed. Athanassios Raftopoulos, 31–47. New York: Nova Science Publishers.

Carrier, Martin. 1994. *The Completeness of Scientific Theories: On the Derivation of Empirical Indicators within a Theoretical Framework – The Case of Physical

Geometry. In *The University of Western Ontario Series in Philosophy of Science*, vol. 53. Dordrecht: Springer.

Chang, Hasok. 2004. *Inventing Temperature: Measurement and Scientific Progress*. In *Oxford Studies in Philosophy of Science*. Oxford: Oxford University Press.

Clark, Andy. 2016. *Surfing Uncertainty: Prediction, Action, and the Embodied Mind*. Oxford: Oxford University Press.

Franklin, Allan. 2015. The Theory-Ladenness of Experiment. *Journal for General Philosophy of Science* 46 (1): 155–166.

Galilei, Galileo. 1989. *Two New Sciences Including Centers of Gravity and Forces of Percussion*. Translated by Stillman Drake. Toronto: Wall & Emerson.

Garber, Daniel. 2013. Remarks on the Pre-history of the Mechanical Philosophy. In *The Mechanization of Natural Philosophy, Boston Studies in the Philosophy* of Science, ed. Daniel Garber and Sophie Roux, vol. 300, 3–26. Dordrecht: Springer.

Gutting, Gary. 1972. Einstein's Discovery of Special Relativity. *Philosophy of Science* 39 (1): 51–68.

Hacking, Ian. 1983. *Representing and Intervening: Introductory Topics in the Philosophy of Natural Science*. Cambridge: Cambridge University Press.

Hanson, Norwood Russell. 1958. *Patterns of Discovery: An Inquiry into the Conceptual Foundations of Science*. Cambridge: Cambridge University Press.

Heidelberger, Michael. 2003. Theory-Ladenness and Scientific Instruments in Experimentation. In *The Philosophy of Scientific Experimentation*, ed. Hans Radder, 138–151. Pittsburgh, PA: University of Pittsburgh Press.

Henry, John. 1986. Occult Qualities and the Experimental Philosophy: Active Principles in pre-Newtonian Matter Theory. *History of Science* 24 (4): 335–381.

Hentschel, Klaus. 1992. Einstein's Attitude towards Experiments: Testing Relativity Theory 1907–1927. *Studies in History and Philosophy of Science Part A* 23 (4): 593–624.

Holton, Gerald. 1996. The Role of *Themata* in Science. *Foundations of Physics* 26 (4): 453–465.

Hoyningen-Huene, Paul. 2013. *Systematicity: The Nature of Science. Oxford Studies in Philosophy of Science*. Oxford: Oxford University Press.

Kitcher, Philip. 1990. The Division of Cognitive Labor. *The Journal of Philosophy* 87 (1): 5–22.

Korbmacher, Max, Flavio Azevedo, Charlotte R. Pennington, et al. 2023. The Replication Crisis has Led to Positive Structural, Procedural, and Community Changes. *Communications Psychology* 1 (1): 1–13.

Krementsov, Nikolai. 1997. *Soviet Science*. Princeton, NJ: Princeton University Press.

Kuhn, Thomas S. 1970. *The Structure of Scientific Revolutions*. In *International Encyclopedia of Unified Science*, 2nd ed. Chicago: University of Chicago Press.

Lagemann, Robert T. 1977. New Light on Old Rays: N-Rays. *American Journal of Physics* 45 (281): 281–284.

Latour, Bruno. 1999. *Pandora's Hope: Essays on the Reality of Science Studies.* Cambridge, MA: Harvard University Press.

Laudan, Larry. 1983. The Demise of the Demarcation Problem. In *Physics, Philosophy and Psychoanalysis: Essays in Honor of Adolf Grünbaum*, Boston Studies in the Philosophy of Science, ed. R. S. Cohen and L. Laudan, vol. 76, 111–127. Dordrecht: D. Reidel.

Lelas, Srđan. 2000. *Science and Modernity: Toward an Integral Theory of Science.* In *Boston Studies in the Philosophy of Science*, vol. 214. Dordrecht: Kluwer.

Levitin, Dimitri. 2022. *The Kingdom of Darkness: Bayle, Newton, and the Emancipation of the European Mind from Philosophy.* Cambridge: Cambridge University Press.

Lindsay, Peter H., and Donald A. Norman. 1977. *Human Information Processing: An Introduction to Psychology.* 2nd ed. New York: Academic Press.

Lyell, Charles. 1835. *Principles of Geology: Being an Inquiry How Far the Former Changes of the Earth's Surface are Referable to Causes Now in Operation.* Vol. 3, 4th edition. London: John Murray.

Mayo, Deborah G. 1994. The New Experimentalism, Topical Hypotheses, and Learning from Error. *PSA: Proceedings of the Biennial Meeting of the Philosophy of Science Association* 1:270–279.

Merton, Robert K. 1973. *The Sociology of Science: Theoretical and Empirical Investigations.* In, ed. Norman W. Storer. Chicago: University of Chicago Press.

Mill, John Stuart. 1882. *A System of Logic, Ratiocinative and Inductive, being a Connected View of the Principles of Evidence, and the Methods of Scientific Investigation.* 8th ed. New York: Harper & Brothers.

Newton, Isaac. 1999. *The Principia: Mathematical Principles of Natural Philosophy.* Translated by I. Bernard Cohen and Anne Whitman. Berkeley, CA: University of California Press.

Nola, Robert, and Howard Sankey. 2000. A Selective Survey of Theories of Scientific Method. In *After Popper, Kuhn and Feyerabend: Recent Issues in Theories of Scientific Method*, Australasian Studies in History and Philosophy of Science, ed. Robert Nola and Howard Sankey, 1–65. Dordrecht: Kluwer.

O'Connor, Cailin, Sanford Goldberg, and Alvin Goldman. 2023. Social Epistemology. In *The Stanford Encyclopedia of Philosophy.* Winter ed., ed. Edward N. Zalta and Uri Nodelman. https://plato.stanford.edu/archives/win2023/entries/epistemology-social/.

Pigliucci, Massimo. 2013. The Demarcation Problem: A (Belated) Response to Laudan. In *Philosophy of Pseudoscience: Reconsidering the Demarcation Problem*, ed. Massimo Pigliucci and Maarten Boudry, 9–28. Chicago: Chicago University Press.

Radder, Hans. 2012. *The Material Realization of Science: From Habermas to Experimentation and Referential Realism.* In *Boston Studies in the Philosophy of Science*, vol. 294. Dordrecht: Springer.

Raftopoulos, Athanassios. 2001. Is Perception Informationally Encapsulated? The Issue of the Theory-ladenness of Perception. *Cognitive Science* 25 (3): 423–451.
Romero, Felipe. 2019. Philosophy of Science and the Replicability Crisis. *Philosophy Compass* 14 (11): 1–14.
Troeltsch, Ernst. 1991. Historical and Dogmatic Method in Theology (1898). Translated by Ephraim Fischoff and revised by Walter Bense. In *Religion in History – Ernst Troeltsch: Essays Translated by James Luther Adams and Walter F. Bense*. Edinburgh: T. & T. Clark.
Wellman, Henry M., and Susan A. Gelman. 1992. Cognitive Development: Foundational Theories of Core Domains. *Annual Review of Psychology* 43 (1): 337–375.
Ziman, John M. 1968. *Public Knowledge: An Essay Concerning the Social Dimension of Science*. Cambridge: Cambridge University Press.

CHAPTER 3

The Roots of Scientific Practice

Abstract On its operational level, modern science is not uniquely Western, for its founders picked up and developed two practices found in the ancient Mediterranean world and transmitted by Muslim scholars: the use of mathematics and the practice of experimentation. But scientific practices have still more ancient roots. They are refined versions of activities that give rise to everyday practical knowledge in all societies. These include observation-based reasoning, counting and measurement, and reliance on the core cognitive categories that all humans share. Modern science transforms these everyday activities. It extends and makes precise the scope of observation by the use of instruments, it standardizes the practice of measurement, and it transforms our core cognitive categories while also guarding against the errors to which they are prone.

Keywords Core cognition • Everyday knowledge • Experiment • Mathematics • Measurement • Standardization

I have distinguished two ways in which science may be said to be 'Western': it could be either Western in origin or Western in character (Sect. 1.3). The present chapter is devoted to the first question: whether science is Western in origin. My argument is that on its operational level (Sect. 2.3.1), the practices that individual scientists engage in have roots that

G. W. Dawes, *Is Science Western?*,
https://doi.org/10.1007/978-3-032-10641-4_3

extend beyond 'the West'. Not only do they have ancient roots, but they are refined versions of practices that all societies share.

3.1 Ancient and Medieval Science

Modern science is indisputably Western in one sense, namely that it began to take on its present forms in seventeenth-century Europe. Historians have traditionally spoken of this development as the 'scientific revolution'. There is a sense in which what happened in Europe at this time *was* revolutionary, for the practice of science was transformed. It was transformed on the level of social practice (Sect. 2.2) through the creation of scientific societies, such as the Academia dei Lincei in Italy (1603), the Royal Society in England (1660), and the Académie des Sciences in France (1666). It was transformed at the level of individual practice (Sect. 2.2) through the merging of the traditions of physics and mathematics and a refinement of the ancient practice of experimentation.

Yet the idea of a scientific revolution is misleading if it is thought to imply a completely new start, a kind of 'year zero' for scientific knowledge. What is often called the 'scientific revolution' would be better called a 'scientific renaissance' (Russo 2004: 2). The ideas and practices that were reborn at this time had a long history, a history that is by no means restricted to the countries we call 'Western'.

Historians often object to the 'Whiggish' practice of studying the past only to seek precursors of the present (Pingree 1992: 561–562). But to ask about the roots of modern scientific practice is a legitimate question, which can stand alongside a study of the past on its own terms. If we ask about the roots of modern science, we find that these can be traced to the ancient Mediterranean world. Before reaching the modern world, these roots pass through medieval societies, in which Arab and Persian scholars had refined the practices that modern science revived. Historically, then, the practices of science are not (merely) Western in origin.

I shall argue for this claim by examining the two sets of practices that are often thought to be distinguishing marks of modern science. The first is the use of mathematics to study the natural world. The second is the practice of experimentation.

3.1.1 Mathematical Sciences

Mathematics did not feature prominently in Aristotle's physics, whose principles were qualitative rather than quantitative. But mathematics played an important role in what Aristotle's medieval followers called 'subalternate sciences': astronomy, optics, harmonics, and mechanics. These were 'subalternate' sciences because they were thought to stand *sub altero* (beneath another). One of the sciences they stood beneath was mathematics. Aristotle himself noted that 'the propositions of optics are subordinate to geometry and those of harmonics to arithmetic' (*APo.* 75b16–17). Practitioners of these subalternate sciences valued the precision that mathematics could provide. The second-century astronomer Claudius Ptolemaeus (Ptolemy) begins his *Almagest* with praise for the precision of mathematical astronomy, which he contrasts with the obscurity of theology and the vagueness of physics (the study of the changeable, sub-lunar world) (Ptolemy 1984: 35–36 [H5–6]).

Were these 'subalternate' sciences Western? If, by a stretch of the imagination, we were to regard ancient Greece as a 'Western' country, then these works could be said to have been written in a Western language. But Greek writers themselves often insisted that the roots of their science lay elsewhere, particularly in Egypt. In the *Timaeus* (22b) Plato recounts a story of an encounter between an Egyptian priest and one of the 'seven sages' of Athenian history, the statesman Solon. The Egyptian priest remarks that 'you Hellenes are never anything but children', since 'there is no old opinion handed down among you by ancient tradition, nor any science which is hoary with age' (Aufrère and Marganne 2019: 502). The story is almost certainly mythical, but could there be any truth in the myth?

There may well be. There are, for instance, ancient reports that Pythagoras (ca. 570–495 BC) spent a long period in Egypt, where he learned geometry. The story is plausible, since there were skilled geometers in Egypt, who applied their art to both the measurement of land and architecture (Aufrère and Marganne 2019: 504). The dependence of Greek science on Egypt is also evident in the most important of the mathematical sciences in antiquity, namely astronomy. Aristotle, for instance, notes his dependence on the astronomical records of the Egyptians (Lloyd 1964: 52). But the Egyptians were, in turn, dependent on the astronomers/astrologers of the late Babylonian empire, for (as one writer puts it) 'the astral science spread from one civilization to another like a highly infectious disease' (Pingree 1992: 556).

The elements that Greek astronomy borrowed from the Babylonians included bodies of astronomical data, the division of the zodiac into twelve signs, the sexagesimal (base-60) system of arithmetic, the use of degrees as units of measurement, the 19-year (Metonic) calendar, and the use of instruments such as the gnomon and clepsydra (water-clock) (Nadal 2016: 398). As this astral science spread, it was altered in the course of its transmission (Pingree 1992: 563). The Greek astronomers, for example, had a theoretical interest—that of constructing geometrical models of heavenly movements—which the Babylonians lacked (Jones 1996: 149). What the Greeks borrowed appears to have been primarily first-order rather than second-order knowledge: that is to say, methods and data rather than theory, in the sense of ways of interpreting that data (Schiefsky 2017: 192).

It remains the case, however, that in the most important of the mathematical sciences of antiquity data and methods were shared, irrespective of where they came from. One scholar speaks of ancient astronomy as 'a continuous tradition in which a change of language was an event of little practical significance' (Jones 1991: 444). Ptolemy, for instance, 'never speaks in national or linguistic terms, but only of sound or unsound deductive methodology' (Jones 1996: 154). Astronomy became even more international when its data and methods travelled through Persia to India. The Indian writer Varāhamihira (*fl.* ca. 550 CE) explicitly recognized the value of Greek astronomy (Neugebauer 1957: 174), while Varāhamihira's work on astrology entered the Muslim world when it was translated into Arabic by al-Bīrūnī (973–1050 CE). Astronomy, the most important of the mathematical sciences of antiquity, was undoubtedly an international affair.

3.1.2 Experimentation

What about the practice of experimentation? It may seem odd to suggest that experimental practice has ancient roots. Since the time of Francis Bacon (1561–1626), the practice of experiment has commonly been regarded as a feature that distinguishes modern science from its predecessors (Lloyd 1964: 50–51). But while a reliance on experiments was a central feature of early modern science (Sect. 2.3.3), it, too, had ancient origins.

Care is needed here because of the various meanings of the term 'experiment' (and its cognate terms in Greek and Latin). A dictionary definition of the corresponding Greek term *peira* is 'trial', 'attempt', or 'experiment', and the word can be used in the sense of a 'test'. But when the medical empiricists of Hellenistic times advocated a reliance on *peirai*, they were referring to 'experiences' rather than controlled experiments (von Staden 1975: 188–190). Similar remarks can be made about the (mostly late) Latin *experimentum*, which can mean 'experience' as well as a 'proof', 'test', 'trial', or 'experiment'. When Roger Bacon (1215–1292). for example, urges his readers to follow the path of *scientia experimentalis*, it would be misleading to translate this as 'experimental science'. 'Experiential knowledge' would be a more accurate rendering. Although Bacon's *scientia experimentalis* did include practical procedures (*experimenta*), such as those used in alchemy, as well as the use of instruments (as in the study of optics and astronomy), it also included simple observation as well as a reliance on second-hand reports.

Within modern science the word 'experiment' is used in a narrower sense. One feature of the modern use has to do with the purpose of experimentation, which is most commonly regarded as that of testing a hypothesis. David Lindberg, for example, writes that an experiment is

> an attempt to confirm or disconfirm a theoretical claim about the nature or behavior of the material world by an observation (under controlled conditions if necessary) made for that purpose, or the gathering of data against which future anticipated theoretical claims may be tested. (Lindberg 2007: 362)

One can certainly find experiments designed to test hypotheses within the practice of ancient science. Astronomical observations, for instance, were made to test claims regarding the structure of the universe.

There is, however, a second feature of the modern understanding of experiment. A practice is thought to be experimental only when involves a manipulation of the conditions under which observations are made (Hansson 2016: 616). But this sense of 'experiment' also has ancient roots. There is a natural affinity, for example, between the ancient science of mechanics and the practice of experimentation. That affinity lies deeper than the trivial fact that mechanical skill is needed to produce experimental apparatus. It lies in the fact that the ancient science of mechanics was the study of simple machines: levers, pulleys, wedges, windlasses, and screws. Machines are artificial contrivances, that produce effects that were thought to be (in the Aristotelian phrase) *para phusin* (beyond nature) (Berryman 2009: 45). They are effects that do not occur in the natural world except when its powers are directed by some human contrivance. It follows that the study of mechanics is the study of effects that can be produced only by manipulating natural conditions, which is precisely what an experiment (in the modern sense) does.

A important figure in the history of ancient mechanics was Hero (or Heron) of Alexandria (*fl.* ca. 60 CE). In the first part of his *Pneumatics* Hero describes an experiment involving a metal sphere designed to illustrate the compressibility of air and, by means of this, Hero's particulate theory of matter (Lloyd 1973: 17–18). Nor was Hero's work unknown to the pioneers of early modern science: it was frequently printed and translated in the sixteenth century and a copy was owned by Galileo (Boas 1949: 42, 48). An even clearer example is found in the work of Philo of Byzantium (ca. 200 BCE), who describes the tests required to improve a catapult, which employ what looks like 'a method of controlled experiment' (Lloyd 1973: 99).

Experiments were not, however, restricted to mechanics. They were undertaken in another of the mathematical sciences, that of optics. Ptolemy, for example, describes experimental apparatus he had devised for measuring angles of incidence and refraction in different media (Lindberg 2007: 108). At least one of his experiments is relatively complex, involving 'the use of three mirrors, one plane, one convex and one concave' (Lloyd 1964: 55), while a second involves the manipulation of a measuring device set up in water (Lloyd 1964: 56).

A third science in which experiments can be found is that of medicine. The author of the (probably Hellenistic) work *On the Heart* describes an experiment performed on a hapless pig to test the idea that a little liquid enters the windpipe when we drink (Lloyd 1964: 67). The anatomist Erasistratus (ca. 304–250 BCE) apparently tested the theory that

> there are invisible effluvia from animals by keeping a bird in a vessel without food for a given period of time and weighing the animal together with the visible excreta and comparing this with its original weight. (Lloyd 1964: 71 n. 1)

But the clearest examples of medical experimentation are found in the work of the physician Claudius Galenus (Galen), a contemporary of Ptolemy, whose writings were to be the primary source of medical knowledge in the medieval Muslim and European worlds.

The best known of Galen's experiments are those he carried out to test his hypotheses about the nervous system. Galen rejected the theory held by thinkers such as the Stoic Chrysippus that 'the ruling part of the soul' (the mind or intellect) was located in the heart, insisting that it was located in the brain and that it transmitted its influence through the nerves of the spinal column. He demonstrated this by a series of experiments performed on living animals. These were an

> extensive series of meticulous vivisections in which the spinal column and nerves are successively severed or blocked at certain spots: the paralysis occurring in the parts below these spots indicates the direction followed by the stimuli as well as their source. (Tieleman 2002: 265)

But Galen also developed what we might call a theory of experimentation. The doctor's craft (*technē*), he argued, involves not just experience but what we may call 'reasoned experience' (Hankinson 2022: 104). The reasoning in question allows the doctor to discern the conditions under which a therapeutic substance will manifest its powers (van der Eijk 2005: 289–290). Galen refers to experience of this kind as 'differentiated experience' (*diōrismenē peira*) (Galen *Op.* 11, p. 685), an idea that resembles our modern conception of a (controlled) experiment (Singer 2022: 185).

Were these ancient experimentalists 'Western'? In many cases, we know little about their origins. The most important centre of Hellenistic science was Alexandria, an Egyptian city where writers in Greek could draw on local traditions of knowledge (Sect. 3.1.1). Yet the question of whether such writers were Western is anachronistic. The ancient Mediterranean world was such a melting pot—a meeting place of ideas from 'Europe, Asia, and Africa' (Sarton 1936: 408)—that it is impossible to draw clear lines between what is culturally 'Eastern' and 'Western'. Galen is a striking example. While he wrote in Greek and worked for some time in Rome, he came from Pergamon, in what is now the north-east of Turkey. George Sarton placed Galen within a long list of those he called 'orientals who took a prominent part in the development of science and philosophy' (Sarton 1936: 429). But to call Galen an 'Oriental' seems as misguided as to call him a 'Westerner'. The labels simply do not fit.

The earliest medieval thinkers to continue this practice of experimentation certainly came from outside what we think of as 'the West'. They were the Arab and Persian natural philosophers who preserved and developed the Greek tradition. Here, too, we find students of optics performing experiments. The Persian mathematician Ibn Sahl (*fl.* ca. 985) used experimental methods to expand on Ptolemy's work regarding angles of reflection and refraction (Lindberg 2007: 184). The leading Arab thinker on optics, Ibn al-Haytham, known in the West as Alhazen (ca. 965–1040), describes many experiments, such as his demonstration that 'the visual object must lie on straight lines drawn from its surface to the "surface of vision"' (which he believed to be the lens of the eye) (Omar 1977: 62). This experiment involved the construction of a simple dioptre (sighting instrument), carefully designed so as to test whether vision could in fact be curvilinear rather than rectilinear (Omar 1977: 63). Another Persian mathematician, Kamāl al-Dīn al-Fārisī (1267–1319), conducted a study of the causes of the rainbow by constructing 'a water-filled glass sphere to simulate a droplet of moisture on which solar rays were allowed to fall' (Lindberg 2007: 184). Arab and Persian thinkers were not alone in his respect. The same experiment was performed in Europe at about the same time by Theodoric of Freiberg (ca. 1250–1311) (Lindberg 2007: 363). But it was non-Europeans who led the way.

It seems, then that from a historical point of view, the phrase 'Western science' embodies a kind of ethnocentrism. It is ethnocentric in its assumption that Europeans alone were responsible for developing the practices it employs (Hess 1995: 63).

3.2 Everyday Practical Knowledge

What I have been arguing is that the practices employed by modern scientists have ancient roots, which extend well beyond the West. But these practices have still more ancient roots. Scientific practices (whether ancient or modern) are refined versions of epistemic practices that are found in all human societies. They are refined version of the activities in which humans have always engaged in the course of gaining what I shall call 'everyday practical knowledge'. To explain what I mean by this phrase, I need to digress for a moment, to say something about the heterogeneity of human societies and the diversity of their practices. So bear with me. I shall return to the sciences shortly.

3.2.1 A Diversity of Epistemic Practices

Every society has a variety of practices, a variety of ways in which individuals interact with their natural and social environment. These practices employ differing ways of representing both the practice itself and its target domain: the features of the world that it seeks to understand and (in some cases) alter. A first category of representations can be described as *religious*. Often enacted in ritual, these make reference to gods, spirits, and ancestors, and recount stories of origin that suggest how one ought to behave. A second category is that of *social representations*, which deal with kinship relations and social roles. (This category commonly overlaps with the religious, since social roles are often grounded in stories of origin, myths that spell out why society has the shape it does.) But a third category of representations—those in which I am interested—stem from those practices in which human beings are 'in most direct contact with nature' (Bloch 1977: 285): the interactions with the environment that provide us with what we need to sustain life, such as food and shelter. Once again, these representations have to do with both the practice itself (craft knowledge) as well as those features of the environment with which the practice interacts. It is this category of representations I am referring to when I speak of *everyday practical knowledge*.

The recognition that all societies have a diversity of epistemic practices dates back to the work of the anthropologist Bronisław Malinowski (1884–1942). Malinowski's predecessor, Lucien Lévy-Bruhl (1857–1939), had drawn a sharp distinction between the *mentalité* of what he called 'primitive' peoples and that of moderns. But Lévy-Bruhl had never lived among the so-called primitives. When Malinowski did so, spending time with the people of the Trobriand Islands off the coast of Papua New Guinea, he discovered that their cultural practices were not all 'cut of the same cloth' (as it were). As Malinowski wrote, such peoples are not continuously 'immersed in a mystical frame of mind' (Malinowski 1925: 28). They have extensive bodies of practical knowledge, whether to do with agriculture, with fishing, or with navigation (Malinowski 1925: 29–34).

Malinowski insisted that all social life has a 'material substratum': the need to sustain human life by interaction with the physical environment (Malinowski 1942: 3). This material substratum is most clearly manifest in the domain of practical knowledge. Because our basic human needs are essentially the same, we find striking commonalities across societies in the ways in which they are met. A good example is provided by tools. As Malinowski wrote,

> it would not be difficult to make an inventory of primitive artifacts: stone implements, fire, dwellings, means of transport and locomotion, and

> weapons-all these exist everywhere. At any level of development the inventory overlaps almost completely from one culture to another. (Malinowski 1942: 3)

It is not just the tools themselves that resemble each other: similar tools give rise to similar rules governing their use and similar forms of cooperative activity (Malinowski 1942: 3).

Everyday practical knowledge is a form of *everyday* knowledge: the 'repertoire of shared and embodied norms, techniques and interpretive frameworks' with which people operate day-by-day (Gardiner 2006: 205). That repertoire includes social representations as well as practical ones. In many societies it will include religious representations, for interaction with gods and ancestors has been (for many peoples) an everyday activity. Social, religious, and practical representations are often interwoven, in the sense of being employed alongside each other, although Malinowski claimed his informants were aware of the differences (Malinowski 1965: 77). Australian aboriginal peoples, for instance, combined practical knowledge of plant propagation with ritual practices to ensure fertility (Sutton and Walshe 2021: 31, 36, 37). While the construction of a Māori meeting house involved sophisticated mechanical techniques for lifting beams (Treadwell 2017: 98), the exercise of this practical knowledge was interwoven with cosmological beliefs (Treadwell 2017: 98). In a similar way, Polynesian navigation was never simply a matter of knowing about star-courses and currents; it also involved appeal to 'ancestral power' (Salmond 2005: 176).

Yet even if the most practical of tools can possess symbolic value (Lefebvre 1987: 8), while using it *as a tool* practitioners employ what I am calling everyday practical knowledge. That knowledge is drawn from observation and trial-and-error reasoning, and is therefore open to revision in the light of new experience. In nineteenth-century Māori society, for example, the practical knowledge employed in building houses changed rapidly, as new techniques and designs became available (Anderson 2015: 140). One prominent *rangatira* (chief) even brought back a prefabricated house from Sydney (Anderson 2015: 141). In a similar way,

> changes inspired by European technology were evident in Māori sailing canoes. A mast fixed with stays, two masts in some canoes, and sails bent onto yards were common in the north, while in the south a large sweep oar,

> the use of oars instead of paddles, and hoisting of squaresails were observed. (Anderson 2015: 140)

For different reasons, Māori religious practices would also change dramatically in the nineteenth century (Sissons 2015: 133). But perhaps the first practices to be altered in the light of new knowledge were those relating to everyday activities such as building a shelter or sailing a boat.

Everyday practical knowledge is culturally transmitted, although it builds on cognitive capacities that appear to be universal and perhaps even innate (Sect. 3.3.3). It is culturally transmitted in the sense that it emerges from activities that need to be learned: how to build a shelter, or plant crops, or sail. None of us is born knowing how to do these things. But since such activities—in varying forms—are common to all societies, the knowledge that arises from them will have features in common. The practices in question are universal practices, giving rise to overlapping ways of representing reality (Sect. 3.3.3).

Here is a simple example. In her study of interactions between eighteenth-century Polynesian and European navigators, Anne Salmond sets out to demonstrate how different their cultural worlds were. But her argument also demonstrates what they had in common. Both knew that one could use the stars to locate one's position at sea, even if they had different ways of doing so (Salmond 2005: 174). Both knew that ocean swells could be an indication of nearby islands, even if the Polynesians attributed the winds and currents to the influence of ancestors (Salmond 2005: 176). When James Cook's Tahitian navigator, Tupaia, 'sent an islander to dive down to check the depth of their ship's rudder' before guiding the *Endeavour* through a passage in a reef (Salmond 2005: 178), he would not have needed to explain what he was doing. For all their differences, European and Polynesian sailors occupied a partially shared world of human practice.

3.2.2 Everyday Knowledge and Science

Everyday practical knowledge may eventually be theorized: practical 'know-how' may be turned into theoretical knowledge, in which its principles are made explicit (Long 1997: 4). Whereas all societies have local knowledge relating to (for example) the planting of crops or the construction of dwellings, not all societies have agricultural and engineering sciences. These sciences are instances of *scientia* (Sect. 1.1): they embody general claims, relating to all kinds of agriculture or building. But what I am focusing on here is the form in which everyday practical knowledge initially exists, as informal, 'folk' or (if more specialized) 'craft' knowledge. Such knowledge arises directly from experience rather than being informed by theory.

To say that it arises from experience is to not say that it is easily gained. Some forms of everyday practical knowledge can require a long period of training. To learn how to sail, for example, requires the mastery of various skills: sensing the direction of the wind, 'reading' the waves, knowing when to haul in and when to ease the sails, and so on. In the ancient Greek work, skilled practical knowledge of this kind was thought of as a form of *mētis*, which can be contrasted with *technē* (Scott 1998: 319–323). In classical and later Greek usage, the two terms are not always opposed. *Technē* is sometimes used in the sense of 'art' or 'skill', making it practically synonymous with *mētis* (Detienne and Vernant 1991: 28, 43). But Aristotle came to use *technē* not merely for practical knowledge—knowledge of the contingencies of everyday life—but for practical knowledge informed by a reasoned account of the activity in question (Parry 2024: Sect. 3). When used in this sense, *technē* is practical skill informed by theory. It is closer to *epistēmē* (the medieval *scientia*) than it is to *mētis*, the latter including practical skills derived immediately from experience.

Even though it has not yet been theorized, skilled everyday practical knowledge (as a form of *mētis*) can be described as 'proto-scientific' (Malinowski 1925: 35). Although religious representations can, and do, change over time, their ultimate principles are presented as handed-down from gods or ancestors and therefore unquestionable (Rappaport 1999: 280). By way of contrast, everyday practical knowledge does not have the same air of sanctity. It is *proto*-scientific because it lacks the reflective self-consciousness that is characteristic of a theorized practice (Sect. 1.1) and because it has not yet been formalized, standardized, and refined (Sect. 3.4). But it is proto-*scientific* insofar as it resembles science. It resembles science in its employment of observation-based reasoning (Sect. 3.3.1), measurement practices (Sect. 3.3.2), and the core knowledge categories that are also used on the operational level of science (Sect. 3.3.3).

Just as Plato disparaged *mētis* in favour of *epistēmē* (Detienne and Vernant 1991: 315–316), the phrase 'proto-science' is sometimes used in a derogatory sense. Thomas Kuhn, for example, contrasts 'proto-sciences' with 'mature sciences', arguing that only the latter permit progress (Kuhn 1970: 244–245). But for the purpose of communal survival and flourishing within a particular environment, proto-scientific knowledge may be more valuable than scientific. As even Aristotle acknowledged (*Met.* 981a11–16), when it comes to practical affairs, knowledge based on experience is more useful than merely theoretical knowledge. Since scientific theories abstract from particular contexts, they are more distant from the concrete realities of human experience (Such 2004: 76). While this gives scientific knowledge greater explanatory power, it also makes it less immediately applicable (Ellen 2003: 50; Atran 2002: 64). Nor should we forget just how much practical knowledge individuals possess in societies that lack a scientific tradition (Sutton and Walshe 2021: 9). Most of us today are comparatively ignorant, for we 'outsource' such knowledge to books, experts, and complex tools. Those who lack books, experts, and complex tools must master this knowledge for themselves. So to call a body of knowledge 'proto-scientific' is not to disparage it. It is merely to indicate what form this knowledge takes.

3.2.3 The Background of Practical Knowledge

I have been arguing that everyday practical knowledge emerges from practices that all societies share. Essential human practices such as growing food, building shelters, and finding one's way across unfamiliar terrain give rise to common ways of representing the environment. But the roots of these commonalities go still deeper, for those practices unfold against a shared background, created by our distinctively human mode of embodiment.

The term 'background' has become something of a technical term. In the work of John Searle, 'the Background' (as he calls it) refers to a set of capacities and assumptions without which we could have no intentional states (such as beliefs and desires). To take one of Searle's examples,

> think of what is necessary, what must be the case, in order that I can now form the intention to go to the refrigerator and get a bottle of cold beer to drink. The biological and cultural resources that I must bring to bear on this task, even to form the intention to perform the task, are (considered in a certain light) truly staggering. But without these resources I could not form the intention at all: standing, walking, opening and closing doors, manipulating bottles, glass, refrigerators, opening, pouring and drinking. (Searle 1983: 143)

Because our social life consists of practices shaped by intentional states, our social life can be seen to unfold against this background of shared capacities and assumptions.

How widely shared are these capacities and assumptions? Searle makes a distinction between two kinds of background. The first is the 'deep Background', which consists of 'capacities that are common to all normal human beings in virtue of their biological makeup', such as 'walking, eating, grasping, perceiving, recognizing, and the pre-intentional stance that takes account of the solidity of things, and the independent existence of objects and other people' (Searle 1983: 143–144). The second is the 'local Background' or 'local cultural practices' which are specific to particular societies (Searle 1983: 144).

The distinction is a useful one, although in practice the two are interwoven. Firstly, the local background seems to be dependent on the universal. Local practices (such as taking a beer from the refrigerator or ordering a coffee) make use of bodily capacities that are common to all people (such as that of walking or speaking). Secondly, the particular way in which the universal background manifests itself is dependent on local factors, for the expression of even universal capacities is culturally shaped. Even so universal a capacity as that of walking is differently expressed in different societies, differing, for example, between France and the United States (Mauss 1973: 72). Our capacity to swim, to take another example, is also differently expressed at different time or in different places (Mauss 1973: 70–71).

Because common human capacities can be exercised differently, we should not give the term 'capacity' too narrow an interpretation, as though it referred to something individuals possess in isolation from their natural and social environments. A capacity is a possibility for an individual only when he or she is located in a particular context (Ingold 2001: 233). My ability to swim, for example, depends not just on my having limbs that can be moved in particular ways; it also depends on my having people around who can instruct me in ways of swimming as well access to a suitable body of water. It is this fact that gives rise to the differences just noted, across time and among different societies. People learning to walk, for example, on surfaces as different as 'and, asphalt, meadow and heath' will learn to walk in subtly different ways (Ingold 2001: 235).

These subtle differences do not, however, undermine the point I am making: that all humans share a deep background of assumptions and capabilities. Differing ways of walking are still recognizably ways of *walking*, just as differing ways of swimming are recognizably ways of *swimming*. Traditional Māori styles of swimming differed in some respects from those practised by Europeans (Haimona 2023: 95). But when Māori women came out to meet early nineteenth-century European ships, the sailors had no hesitation in describing what they were doing as swimming (Anderson 2015: 141). These shared background assumptions and capacities arise from the fact that humans have a shared mode of embodiment, with its distinctive cycle of life, lived out in physical and social environments that have features in common (Moran 2011: 487).

3.3 Three Shared Features

We have just seen that everyday practical knowledge draws from a background of shared human capacities and assumptions. But the operational level of scientific practice also draws from that background, even if it refines the capacities in question (Sect. 3.4). To put this more simply, everyday knowledge and scientific practice have shared features (Johnson-Laird 2006: 421–422). The three shared features on which I shall focus are (a) observation-based reasoning, (b) practices of counting and measurement, and (c) the cognitive categories that feature in what is often called 'core cognition'.

3.3.1 Observation-Based Reasoning

It is common to remark that the modern sciences are evidence-based, but this does not go far enough. The sciences base their conclusions on a very particular kind of evidence. In the first instance, much of it must be (in practice) the evidence of testimony. I read about an experiment (and if it is published in a peer-reviewed journal) I will normally accept the reported outcome on the testimony of those who claim to have performed it. But while testimony is important, modern science does not regard the testimony of others as sufficient (Sect. 4.3.2). Central to the idea of an experiment is that it is replicable: its outcome can be checked by performing the same procedure.

Do all societies employ reasoning based on and controlled by what can be observed? There are good *a priori* reasons to think they would. As one author writes,

> no society could survive for any length of time without conducting a large part of its daily activities by the principle of belief according to the

> [observable] evidence. You cannot farm [for example] without some rationally based knowledge of soils and seeds and of meteorology. (Wiredu 1997: 322)

But we do not need to rely on *a priori* arguments. Once again following in the footsteps of Malinowski (Sect. 3.2.1), anthropologists have offered examples of careful observation-based reasoning even in small-scale societies that lack a scientific tradition.

A first such example can be found among the agricultural practices of peasant farmers in Fujian Province in China. Paul Chandler has documented not merely the rich ecological knowledge these farmers have, but also the reasoning in which they engage with their environment, reasoning that is (he writes) 'functionally similar' to that employed by scientists (Chandler 1991: 63). While these peasant farmers do not engage in controlled experiments and have no access to scientific literature, they successfully apply 'logic and deductive reasoning' to observations accumulated over a number of generations and transmitted 'by word of mouth alone' (Chandler 1991: 65). Insofar as this process does span the generations, we can speak here not just of proto-science, but of a proto-scientific tradition.

The Chinese peasants Chandler studied were an agricultural society, but Nicholas Blurton-Jones and Melvin Konner observed similarly proto-scientific reasoning among a hunting and gathering people, namely the !Kung bushmen of the Kalahari desert. When tracking an animal. Blurton-Jones and Konner write, the !Kung are continually forming and revising hypotheses about the location and condition of the animal they are following, hypotheses that can be discussed with fellow hunters as new evidence emerges (Blurton-Jones and Konner 1976: 342–343). When they engage in such discussions, they are careful to distinguish observed data from mere hearsay and to distinguish data from explanation (Blurton-Jones and Konner 1976: 328). They also distinguish causation from mere correlation. Take, for instance, a reported discussion about the birds called 'honey guides', so-called because they lead humans to beehives, so they can feast on the grubs and wax that are left behind after the honey has been extracted. As Blurton-Jones and Konner write, one man

> suggested that the honey guide sometimes leads leopards to honey. This suggestion was followed by the objective statement that if you are following a bird, you sometimes see a leopard. The statement was taken up immediately

and negated by someone who claimed that the bird leads people to leopards, not leopards to the honey. (Blurton-Jones and Konner 1976: 329)

What we see here are differing possible explanations of a phenomenon being debated.

In such contexts the !Kung hunters are extending their knowledge by discussion. I have already noted the importance of the social dimension of the sciences (Sect. 3.4.3): the collective activities scientists engage in (Sect. 2.2). But this example reminds us that these collective practices almost certainly have ancient roots, being found (as far as we can tell) in all human societies (Mercier 2011: 92). These discussions employ causal reasoning of a kind resembling that used by modern scientists (Boehm 1996: 772, 786). It is often a form of 'inference to the best explanation' or (to use C. S. Peirce's term) 'abduction', which passes from observed phenomena to unobserved causes or states of affairs (Carruthers 2002: 78, 79).

3.3.2 Counting and Measurement

There is a particular form of observation-based reasoning that plays a central role in modern science, namely the practice of counting and measurement (Stevens 1958: 385). But counting and measurement are also everyday practices, found in all societies. They, too, are practices that modern science refines rather than invents (Sect. 3.4.1).

The relation between counting and measurement is a contested one. Counting involves recognizing sets of individual objects and of mapping them onto either number words or other objects (such as sticks). If we define measurement as simply 'the pinning of numbers on things' (Stevens 1958: 384), then counting could be considered a form of measurement. By this definition, measurement (in the form of counting) would a practice common to all humans; in the case of stick counting, it may even predate modern humans (Clawson 1994: 24).

One can argue, however, that this is too liberal a definition of 'measurement'. The word should arguably be used only for the assignment of numbers to continuous properties rather than discrete objects (Berka 1983: 106–110). But even on this narrower definition, the practice of measurement appears to be universal. It may have gained added significance with the agricultural revolution that occurred about 13,000 years ago. In agricultural societies fields must be measured to ascertain their area and surplus grain needs to be measured by weight or volume in order to be stored

(Clawson 1994: 50–51). But units of measurement—particularly body-based units of measurement—have been documented in all societies, even those lacking settled agricultural practices (Cooperrider and Gentner 2019: 3–6).

Some of the activities that could be described as measurement do not involve numbers. Ancient kiln-operators, for instance, judged the intensity of heat inside their kilns using test-pieces of clay, which gave off different colours at differing temperatures (McCaskey 2020: 407). Can we describe this as 'measuring' the heat of the kiln? I believe we can. One might argue that the term 'measurement' should be used only of activities that result in accurate readings (Mari et al. 2017: 217) and modern science has greatly increased the accuracy of our measuring techniques (Sect. 3.4.2). But estimating the heat of a kiln by the colours of its contents can be done with a degree of accuracy that is more than sufficient for the task at hand (Noble 1960: 311).

3.3.3 Core Cognitive Competencies

There is a third feature of everyday practical knowledge which continues to influence scientific practice. It is the use of the categories employed in what I shall call (following Susan Carey) 'core cognition': a set of cognitive competencies that all humans share (Carey 2009: 14).

The idea that all humans share a set of cognitive competencies remains controversial. In particular, there are debates about which competencies fall into this category (Eraña and Martínez 2004: 706) and to what extent they are innate (Spelke 1998), rather than developing very early (Karmiloff-Smith 1994). But it does appear that 'there are some abilities, concepts, and beliefs' that all human beings easily acquire, which 'are at the core of our reasoning processes' (Eraña and Martínez 2004: 722). I have already mentioned some of the general principles that shape visual processing (Sect. 2.3.2). But these shared abilities, concepts, and beliefs have a wider role, operating within five domains of knowledge. Four of these have to do with

> [i] inanimate objects and their mechanical interactions, [ii] agents and their goal-directed actions, [iii] sets and their numerical relationships of ordering, addition and subtraction, and [iv] places in their spatial layout and their geometric relationships. (Spelke and Kinzler 2007: 89)

We can add to this list (v) the domain of plants and (non-human) animals (Atran 2002: 50–55).

While cultural influences build on these core competencies, the competencies themselves develop in all humans, being activated by similar bodily experiences (Baillargeon 1998: 510–511, 515–516). Indeed in some cases they appear to pre-date humans, being found in other primates (Carey 2009: 88–90). It follows that from an early age all humans develop similar ways of representing objects, agents, numbers, space, and living things. Core cognitive abilities do not merely shape childhood learning; they influence how we continue to perceive the natural and social world (Carey 2009: 69–87), being particularly important in the realm I am calling 'everyday practical knowledge'.

The anthropologist Robin Horton described these shared ways of representing the physical and social worlds as 'primary theory', which he distinguished from the 'secondary theories' that are created to explain phenomena (Horton 1993: 320–321). It may be better to avoid the word 'theory' in this context, since core cognition arguably does not have the same form or function as theoretical knowledge (Carey 2009: 363). But Horton was right to note that this kind of knowledge 'does not differ very much from community to community or from culture to culture' (Horton 1993: 321). As part of the deep background of human activity (Sect. 3.2.3). it emerges from a shared mode of embodiment: 'our built-in systems of perception and motion' (Peruzzi 1994: 198).

Core cognitive categories are altered by learning, and in particular by the cultural representations we inherit. Those cultural representations sometimes enrich, sometimes violate, and sometimes replace our core cognitive concepts (Boyer 1998: 879–882). Modern science appears (at least at first sight) to have replaced our core categories. Many of its theories are 'counter-intuitive', in the sense of violating the assumptions that we are born predisposed to make. Aristotelian theories of motion, for example, employed ideas that we naturally find intuitive (McCloskey 1983: 123). But from the time of Galileo's rejection of Aristotelian theories, the view of the world offered by modern physics has become increasingly strange, by everyday standards. While modern biology has remained more closely linked to the everyday world, it has also revised many of our 'folk-biological' intuitions (Atran 2002: 42). Folk biology, for instance, treats each species as having a 'unique causal essence, or inherent underlying nature' (Atran 2002: 63). But biologists no longer think about species in this 'essentialized' manner (Ereshefsky 2022: Sect. 2.1).

Yet this not mean that science has simply abandoned the categories of core cognition. Although the concepts employed in scientific theories may differ from everyday ones, scientific theorizing remains influenced by our core cognitive categories. That influence is not always positive. Intuitive categories can distort scientific thinking when the latter is 'not underpinned by an explicit theoretical framework' (De Cruz and De Smedt 2007: 365). But those core categories also play positive roles within scientific practice.

A first of these is that even our scientific concepts are created with the aid of our intuitive ones. The process by which we create new concepts is described by Susan Carey as 'Quinean bootstrapping' (Carey 2009: 33), after the philosopher W. v. O. Quine (1908–2000). The process begins with learning scientific symbols and their relations to each other. These symbols initially act as 'placeholders, at most only partially interpreted with respect to antecedent concepts' (Carey 2009: 307). We then more fully interpret the scientific symbols by analogical reasoning, abduction, thought experiment, and induction (Carey 2009: 307). In the course of this process, our core cognitive categories form the 'building blocks' with which the new concepts are developed (Baillargeon and Carey 2012: 59).

Carey's 'bootstrapping' account of the formation of scientific theories continues to be debated (Beck 2017: 112–113). But even leaving her view aside, we can see our core cognitive competencies at work on the operational level of scientific practice. The core categories of human thought play a role in shaping the ways in which scientists divide up phenomena and search for regularities among them. Take, once again, biological science. Among human classifications of plants and animals, for example, folk species 'often correspond to scientific genera (e.g. *oak*) or species (e.g. *dog*)', at least for organisms commonly encountered, 'such as larger vertebrates and flowering plants' (Atran 2002: 46). These shared categories—shared, that is, by both 'the folk' and professional biologists—are employed in the search for common properties (Atran 2002: 44). Even in modern biology folk categories continue to have heuristic value (Atran 2002: 66) and remain, in practice, indispensable for scientific work (Atran 2002: 72).

Similar remarks may be made regarding the experimental sciences. Here, too, assumptions derived from our core cognitive competencies are employed on the operational level of scientific practice. As Travis Norsen writes,

> in a typical experiment, the outcome will be registered in the position (or some other perceptually obvious feature) of a macroscopic object – e.g., the position, along some scale, of a pointer or 'needle', the color of some flashing light, or a number projected on a computer screen or printed on a sheet of paper. (Norsen 2007: 323)

These descriptions of experimental observations are 'expressed in common language, suitably refined by the vocabulary of classical physics' (Bohr 1963: 91). This use of core concepts is essential if the sciences are to remain connected with the phenomenal reality of everyday human experience (Atran 2002: 67).

I have highlighted three features that science and everyday knowledge have in common. The commonality is unsurprising, since all three features form part of the deep background of human life (Sect. 3.2.3). Observation-based reasoning relies on induction—the practice of generalizing from particular instances—that depends on a background assumption of regularity in the natural world (Searle 1983: 133). The earliest known practices of counting and measurement use body-based units that are strikingly similar across cultures, and whose usage persists even when more abstract and standard measures are developed (Kaaronen et al. 2023: 948). Finally, what psychologists have come to call our 'core' cognitive competencies are the capabilities all humans develop, very early in their lives, to categorize and reason about their physical and social world.

3.4 Refining Everyday Practices

Scientific practices are not, however, identical with those found in everyday life, for the sciences also extend and transform our everyday practices and categories. They formalize our everyday reasoning, extend and standardize our practices of observation and measurement, and revise our core cognitive categories.

3.4.1 Reasoning Formalized

I have argued for the use of observation-based reasoning in all societies, where it is employed in the course of obtaining everyday practical knowledge (Sect. 3.3.1). But the everyday reasoning of individuals is not, for the most part, formal reasoning, in the sense of reasoning that involves propositions linked by inferences. First of all, everyday reasoning involves the choice of goals as well as decisions among competing ways of achieving those goals. But it is also different in form from the reasoning studied by logicians. Everyday reasoning involves mental models: structures of closely linked representations that may involve language (Bloch 1998: 25), but which need not do so (Johnson-Laird 2006: 182). Those mental models that arise from bodily interaction with the physical environment seem to shape much of our thinking, even when that thinking is formalized (Peruzzi 1994: 195, 205–218). Our everyday employment of such models occurs (for the most part) beneath the level of conscious awareness (Johnson-Laird 2006: 61), rather than on the basis of formal, rule-based reasoning. Indeed formal reasoning would be too slow to motivate everyday activity: structured models allow for much faster responses (Bloch 1998: 24–25). So while logic has a role to play in everyday reasoning, it is not logic that gives it its shape.

If, however, we are to evaluate everyday reasoning, we need to engage in metacognition: thinking about our thinking (Johnson-Laird 2010: 18247). We need to formalize it, setting out the assumptions on which it appears to be based. A first step in this process occurs when we reflect on what we were thinking and try to express it in words, a highly fallible process, since it will often involve *post hoc* rationalization (Bloch 1998: 26). But the sciences engage in a further level of formalization. Being practices that employ systematic second-order reflection (Sect. 1.1), they make explicit and assess not just our reasoning in particular cases, but the rules that underlie all of our reasoning (insofar as that reasoning is truth-tracking).

This process began with the creation of a general science of logic: a systematic account of how truth is preserved in arguments. Aristotle's logical writings, collectively known as the *Organon*, were particularly influential in this regard. Some of these—his *Prior* and *Posterior Analytics*—were devoted to scientific reasoning: Francis Bacon (1561–1626) was offering an alternative account when he published his *Novum Organum*. In our own time such second-order reflection is more commonly carried out by philosophers than by scientists. But scientists have also engaged in this activity. Newton set out what he called *regulae philosophandi* (rules for doing [natural] philosophy) in book three of his *Principia*. William Whewell (1794–1866) modified Newton's principles, while engaging in a debate with John Stuart Mill. Such rules offer scientists principles against which to assess their own reasoning and that of their peers.

3.4.2 Practices Extended and Standardized

I have spoken of 'observation-based reasoning' as an everyday practice on which science builds. But the scientific practice of observation goes beyond the everyday one, extending its powers of by means of instruments. So in the case of the sciences it is better to speak of the broader category of the 'empirically detectable', where 'detection' is 'observation enhanced by our best scientific instruments' (van der Merwe 2024, Sect. 5.1).

The use of instruments was already an established practice in the first mathematical science (Lindberg 2007: 362), that of astronomy (Sect. 3.1.1). The earliest instruments did not extend everyday perception, since astronomers still used the naked eye. But they enabled everyday perceptual practices to yield more precise measurements. The simplest such instrument was the gnomon (used to calculate the angle and orientation of a shadow cast by the sun) while the most complex was the astrolabe (a physical model of the relative positions of heavenly bodies, with moving parts to allow for calculations). It was only with the invention of the telescope and microscope in the seventeenth century that everyday observation practices were extended, to enable one to see what could not be seen with the eyes alone.

I have just suggested that the earliest scientific instruments made observation more precise by way of measurement. The practice of measurement is itself an everyday practice (Sect. 3.3.2), but it can contribute to scientific knowledge only if the measures being employed are standardized. Without standardization there could be no reliable testing of scientific results by other practitioners. Take, for example, the measurement of temperature. While the practice of judging levels of heat and cold is presumably common to all societies (Sect. 3.3.2), this practice became of scientific utility only when there existed reliable instruments using standard scales (McCaskey 2020: 415).

For a long time, there were no standard scales for measuring not only temperature, but also length, weight, and volume. In late medieval Europe there was 'an incoherent multitude of systems of measurement that varied from city to city and village to village' (Febvre 1982: 389). Galileo was forced to invent his own units of measurement to deal with very small distances and short periods of time (Wootton 2010: 48). When he used locally accepted measures, it was unclear to others what they referred to. In attempting to understand Galileo's work on the speed of falling bodies, Marin Mersenne had to purchase a Florentine measuring stick to understand what was meant by a *braccio* (Wootton 2010: 49). Even then, he may have been in error, for Galileo may have been using the Venetian rather than the Florentine *braccio* (Wootton 2010: 49 n. 21).

The standardization of measures involved making them more abstract: less tied to tangible objects and not restricted to particular contexts of use (Cooperrider and Gentner 2019: 1). This process had its roots outside of science. In Europe it began as a result of the commercial revolution of late medieval times (Sect. 4.2.2), while in China it began much earlier, driven by 'legalist' ideas about how to govern the state (Bodde 1991: 136). But however it began, standardization was vital for the development of science. The process took some time. It was only in the late 1700s that scientific measures began to be standardized worldwide, greatly assisted by the invention of the metric system in France (Treese 2018:11). This process went hand-in-hand with the standardization of scientific equipment. 'The first piece of experimental apparatus' to become 'standardized and widely available' was arguably the barometer invented by Galileo's disciple Evangelista Torricelli in 1643 (Wootton 2015: 340).

3.4.3 Core Cognition Criticized

I have argued that the sciences employ the same concepts and competencies that that all humans share, particularly at the operational level, and that they must do so in order to relate their theories to everyday experience and practice (Sect. 3.3.3). I have also noted that the sciences also alter those concepts when constructing theories, with the result that many scientific theories appear counter-intuitive (Sect. 3.3.3). What I want to note here are the scientific practices designed to overcome the fallacies to which everyday thinking is prone.

It is reasonable to assume that our core cognitive competences have evolved because they are a more or less reliable means of acquiring everyday practical knowledge. But everyday thinking is far from perfectly rational. All humans are predisposed to make certain kinds of error (Johnson-Laird 2010: 18249), a fact highlighted by the 'heuristics and biases' literature (Tversky and Kahneman 1982: 4–11). We 'operate with vague hypotheses, perform fallacious inferences, have memory lapses, and display confirmation bias' (McCauley 2000: 67). In making judgements about whether a phenomenon belongs to a certain class, we overlook prior probabilities or 'base-rate frequencies'. (We may, for instance, estimate the likelihood an individual has a particular occupation in a way that overlooks the distribution of occupations in the population as a whole.) Similarly, we overlook the significance of sample size, make judgements based on samples that are more readily available rather than truly representative, and misunderstand the nature of chance events. Even 'experienced researchers are ... prone to the same biases – when they think intuitively' (Tversky and Kahneman 1982: 18).

The development of rules of scientific reasoning is one way in which the sciences have tried to avoid these common fallacies (Sect. 3.4.1). Those rules include the principles of probabilistic reasoning, which is arguably the least 'natural' form of thinking we can employ (Elvin 2010: 401). But there is a still more important way in which the sciences have tried to counter our everyday tendencies to error. It is the development of a set of collective practices (Sect. 2.2), the best known of which is peer review. Particularly if the community is large enough and encourages a variety of views, such collective practices can help counter our individual cognitive biases (De Cruz and De Smedt 2012: 427).

In one sense there is nothing uniquely Western about such collective practices, for all societies engage in collective reasoning in order to solve practical problems (Sect. 3.3.1). But what is more distinctively Western is the way in which modern science has institutionalized this practice, by creating communities of researchers enjoying a partial autonomy from the surrounding society (Sect. 4.3.2). Within these communities collective criticism can counteract the biases to which we are prone when thinking intuitively.

Science does, then, build on everyday epistemic practices. It builds on our everyday practices of observation-based reasoning and counting and measurement, as well as extending (on the theoretical level) and continuing to employ (on the operational level) the core cognitive categories that all humans share. But science does not leave these everyday practices unchanged: it formalizes, extends, standardizes, and refines them so as to make them more reliable.

References

Anderson, Atholl. 2015. Te Ao Tawhito: The Old World. In *Tangata Whenua: A History*, ed. Atholl Anderson, Judith Binney, and Aroha Harris, 9–163. Wellington: Bridget Williams Books.

Atran, Scott. 2002. Modular and Cultural Factors in Biological Understanding: An Experimental Approach to the Cognitive Basis of Science. In *The Cognitive*

Basis of Science, ed. Peter Carruthers, Stephen Stich, and Michael Siegal, 41–72. Cambridge: Cambridge University Press.

Aufrère, Sydney H., and Marie-Hélène Marganne. 2019. Encounters Between Greek and Egyptian Science. In *A Companion to Greco-Roman and Late Antique Egypt*, Blackwell Companions to the Ancient World, ed. Katelijn Vandorpe, 501–518. Hoboken, NJ: Wiley Blackwell.

Baillargeon, Renée. 1998. Infants' Understanding of the Physical World. In *Advances in Psychological Science, Vol. 2: Biological and Cognitive Aspects*, ed. Michel Sabourin, Fergus Craik, and Michèle Robert, 503–529. East Sussex, UK: Psychology Press.

Baillargeon, Renée, and Susan Carey. 2012. Core Cognition and Beyond: The Acquisition of Physical and Numerical Knowledge. In *Early Childhood Development and Later Outcome*, ed. Sabina M. Pauen, 33–65. Cambridge: Cambridge University Press.

Beck, Jacob. 2017. Can Bootstrapping Explain Concept Learning? *Cognition* 158:110–121.

Berka, Karel. 1983. *Measurement: Its Concepts, Theories and Problems.* Translated by Augustin Riska. Boston Studies in the Philosophy of Science 72. Dordrecht: D. Reidel.

Berryman, Sylvia. 2009. *The Mechanical Hypothesis in Ancient Greek Natural Philosophy.* Cambridge: Cambridge University Press.

Bloch, Maurice. 1977. The Past and the Present in the Present. *Man NS* 12:278–292.

Bloch, Maurice. 1998. *How We Think They Think: Anthropological Approaches to Cognition, Memory, and Literacy.* Boulder, CO: Westview Press.

Blurton-Jones, Nicholas, and Melvin J. Konner. 1976. !Kung Knowledge of Animal Behavior. In *Kalahari Hunter-Gatherers: Studies of the !Kung San and Their Neighbours*, ed. Richard B. Lee and Irven DeVore, 325–348. Cambridge, MA: Harvard University Press.

Boas, Marie. 1949. Hero's Pneumatica: A Study of Its Transmission and Influence. *Isis* 40 (1): 38–48.

Bodde, Derk. 1991. *Chinese Thought, Society, and Science: The Intellectual and Social Background of Science and Technology in Pre-modern China.* Honolulu, HI: University of Hawaii Press.

Boehm, Christopher. 1996. Emergency Decisions, Cultural-Selection Mechanics, and Group Selection. *Current Anthropology* 37 (5): 763–793.

Bohr, Niels. 1963. *Essays 1958–1962 on Atomic Physics and Human Knowledge.* New York: Interscience Publishers.

Boyer, Pascal. 1998. Cognitive Tracks of Cultural Inheritance: How Evolved Intuitive Ontology Governs Cultural Transmission. *American Anthropologist* 100 (4): 876–889.

Carey, Susan. 2009. *The Origin of Concepts. Oxford Studies in Cognitive Development.* Oxford: Oxford University Press.

Carruthers, Peter. 2002. The Roots of Scientific Reasoning: Infancy, Modularity and the Art of Tracking. In *The Cognitive Basis of Science*, ed. Peter Carruthers, Stephen Stich, and Michael Siegal, 73–95. Cambridge: Cambridge University Press.

Chandler, Paul M. 1991. The Indigenous Knowledge of Ecological Processes among Peasants in the People's Republic of China. *Agriculture and Human Values* 8 (1): 59–66.

Clawson, Calvin C. 1994. *The Mathematical Traveller: Exploring the Grand History of Numbers.* Dordrecht: Springer.

Cooperrider, Kensy, and Dedre Gentner. 2019. The Career of Measurement. *Cognition* 191 (2019): 103942.

De Cruz, Helen, and Johan De Smedt. 2007. The Role of Intuitive Ontologies in Scientific Understanding – The Case of Human Evolution. *Biology and Philosophy* 22:351–368.

De Cruz, Helen, and Johan De Smedt. 2012. Evolved Cognitive Biases and the Epistemic Status of Scientific Beliefs. *Philosophical Studies* 157:411–429.

Detienne, Marcel, and Jean-Pierre Vernant. 1991. *Cunning Intelligence in Greek Culture and Society* (1974). Translated by Janet Lloyd. Chicago: Chicago University Press.

Ellen, Roy. 2003. Arbitrariness and Necessity in Ethnobiological Classification: Notes on Some Persisting Issues. In *Nature Knowledge: Ethnoscience, Cognition, and Utility*, ed. Glauco Sanga and Gherardo Ortalli, 47–56. New York: Berghahn Books.

Elvin, Mark. 2010. Personal Luck: Why Premodern China – Probably – did not Develop Probabilistic Thinking. In *Concepts of Nature: A Chinese-European Cross-Cultural Perspective*, Conceptual History and Chinese Linguistics, ed. Hans Ulrich Vogel and Günter Dux, vol. 1, 400–468. Leiden: E. J. Brill.

Eraña, Ángeles, and Sergio Martínez. 2004. The Heuristic Structure of Scientific Knowledge. *Journal of Cognition and Culture* 4 (3–4): 701–729.

Ereshefsky, Marc. 2022. Species. In *The Stanford Encyclopedia of Philosophy.* Summer ed., ed. Edward N. Zalta https://plato.stanford.edu/archives/sum2022/entries/species/.

Febvre, Lucien. 1982. *The Problem of Unbelief in the Sixteenth Century: The Religion of Rabelais, translated by Beatrice Gottlieb.* Cambridge, MA: Harvard University Press.

Gardiner, Michael E. 2006. Everyday Knowledge. *Theory, Culture and Society* 23 (2–3): 205–207.

Haimona, Mark Waihape. 2023. *He oranga tō te wai.* Māori Pedagogies: A Culturally Responsive View of Aquatic Education and Drowning Prevention in Aotearoa. Masters Thesis. University of Waikato.

Hankinson, R. J. 2022. Discovery, Method, and Justification: Galen and the Determination of Therapy. In *Galen's Epistemology: Experience, Reason, and Method in Ancient Medicine*, ed. R. J. Hankinson and Matyáš Havrda, 79–115. Cambridge: Cambridge University Press.

Hansson, Sven Ove. 2016. Experiments: Why and How? *Science and Engineering Ethics* 22 (3): 613–632.

Hess, David J. 1995. *Science and Technology in a Multicultural World: The Cultural Politics of Facts and Artifacts.* New York: Columbia University Press.

Horton, Robin. 1993. *Patterns of Thought in Africa and the West: Essays on Magic, Religion and Science.* Cambridge: Cambridge University Press.

Ingold, Tim. 2001. Evolving Skills. In *Alas, Poor Darwin: Arguments Against Evolutionary Psychology*, ed. Hilary Rose and Steven Rose, 225–246. London: Vintage.

Johnson-Laird, Philip N. 2006. *How We Reason.* Oxford: Oxford University Press.

Johnson-Laird, Philip N. 2010. Mental Models and Human Reasoning. *Proceedings of the National Academy of Sciences* 107 (43): 18243–18250.

Jones, Alexander. 1991. The Adaptation of Babylonian Methods in Greek Numerical Astronomy. *Isis* 82 (3): 440–453.

Jones, Alexander. 1996. On Babylonian Astronomy and its Greek Metamorphoses. In *Tradition, Transmission, Transformation: Proceedings of Two Conferences on Pre-Modern Science Held at the University of Oklahoma*, Collection de Travaux de L'Academie Internationale d'Histoire des Sciences, ed. F. Jamil Ragep and Sally P. Ragep, vol. 37, 139–155. Leiden: E. J. Brill.

Kaaronen, Roope O., Mikael A. Manninen, and Jussi T. Eronen. 2023. Body-based Units of Measure in Cultural Evolution. *Science* 380 (6648): 948–954.

Karmiloff-Smith, Annette. 1994. Précis of Beyond Modularity: A Developmental Perspective on Cognitive Science. *Behavioral and Brain Sciences* 17 (4): 693–707.

Kuhn, Thomas S. 1970. *The Structure of Scientific Revolutions.* In *International Encyclopedia of Unified Science*, 2nd ed. Chicago: University of Chicago Press.

Lefebvre, Henri. 1987. The Everyday and Everydayness. *Yale French Studies* 73):7–11.

Lindberg, David C. 2007. *The Beginnings of Western Science: The European Scientific Tradition in Philosophical, Religious, and Institutional Context, Prehistory to A.D. 1450.* 2nd ed. Chicago: University of Chicago Press.

Lloyd, G. E. R. 1964. Experiment in Early Greek Philosophy and Medicine. *Proceedings of the Cambridge Philological Society NS* 10:50–72.

Lloyd, G. E. R. 1973. *Greek Science After Aristotle. Ancient Culture and Society.* New York: W. W. Norton & Company.

Long, Pamela O. 1997. Power, Patronage, and the Authorship of *Ars*: From Mechanical Know-How to Mechanical Knowledge in the Last Scribal Age. *Isis* 88 (1): 1–41.

Malinowski, Bronisław. 1925. Magic, Science and Religion. In *Science, Religion and Reality*, 20–84. London: Sheldon Press.

Malinowski, Bronisław. 1942. A New Instrument for the Interpretation of Law – Especially Primitive. *Lawyers Guild Review* 2 (3): 1–12.

Malinowski, Bronisław. 1965. *Coral Gardens and Their Magic.* In *Indiana University Studies in the History and Theory of Linguistics*, vol. 1 (1935). Bloomington, IN: Indiana University Press.

Mari, Luca, Andrew Maul, David Torres Irribarra, and Mark Wilson. 2017. Quantities, Quantification, and the Necessary and Sufficient Conditions for Measurement. *Measurement* 100:115–121.

Mauss, Marcel. 1973. Techniques of the Body (1935). *Economy and Society* 2 (1): 70–88.

McCaskey, John P. 2020. History of "Temperature": Maturation of a Measurement Concept. *Annals of Science* 77 (4): 399–444.

McCauley, Robert N. 2000. The Naturalness of Religion and the Unnaturalness of Science. In *Explanation and Cognition*, ed. Frank C. Keil and Robert A. Wilson, 61–85. Cambridge, MA: MIT Press.

McCloskey, Michael. 1983. Intuitive Physics. *Scientific American* 248 (4): 122–131.

Mercier, Hugo. 2011. On the Universality of Argumentative Reasoning. *Journal of Cognition and Culture* 11 (1–2): 85–113.

Moran, Dermot. 2011. "Even the Papuan is a Man and not a Beast": Husserl on Universalism and the Relativity of Cultures. *Journal of the History of Philosophy* 49 (4): 463–494.

Nadal, Robert. 2016. Les Lois. L'Astronomie: Observer et Calculer. In *L'Encyclopédie du Ciel*, Bouquins, ed. D' Arnaud Zucker, 387–775. Paris: Éditions Robert Laffont.

Neugebauer, Otto. 1957. *The Exact Sciences in Antiquity.* 2nd ed. New York: Dover Publications.

Noble, Joseph V. 1960. The Technique of Attic Vase-Painting. *American Journal of Archaeology* 64 (4): 307–318.

Norsen, Travis. 2007. Against "Realism". *Foundations of Physics* 37 (3): 311–340.

Omar, Saleh Beshara. 1977. *Ibn al-Haytham's Optics: A Study of the Origins of Experimental Science.* Minneapolis: Bibliotheca Islamica.

Parry, Richard. 2024. Episteme *and* Techne. In *Stanford Encyclopedia of Philosophy*, Spring ed., ed. Edward N. Zalta and Uri Nodelman. https://plato-stanford-edu.ezproxy.otago.ac.nz/archives/spr2024/entries/episteme-techne/.

Peruzzi, Alberto. 1994. An Essay on the Notion of Schema. In *Shapes of Forms: From Gestalt Psychology and Phenomenology to Ontology and Mathematics*, Synthese Library, ed. Liliana Albertazzi, vol. 275, 191–243. New York: Oxford University Press.

Pingree, David. 1992. Hellenophilia versus the History of Science. *Isis* 83 (4): 554–563.

Ptolemy, Claudius. 1984. *Ptolemy's Almagest*. Translated by G. J. Toomer. London: Duckworth.

Rappaport, Roy A. 1999. *Ritual and Religion in the Making of Humanity*. In *Cambridge Studies in Social and Cultural Anthropology*, vol. 110. Cambridge: Cambridge University Press.

Russo, Lucio. 2004. *The Forgotten Revolution: How Science Was Born in 300* BC *and Why It Had to Be Reborn*. Translated by Silvio Levy. Berlin: Springer.

Salmond, Anne. 2005. Their Body is Different, Our Body is Different: European and Tahitian Navigators in the 18th Century. *History and Anthropology* 16 (2): 167–186.

Sarton, George. 1936. The Unity and Diversity of the Mediterranean World. *Osiris* 2:406–463.

Schiefsky, Mark. 2017. The Creation of Second-Order Knowledge in Ancient Greek Science as a Process in the Globalization of Knowledge. In *The Globalization of Knowledge in History*, ed. Jürgen Renn, 191–202. Berlin: Edition Open Access.

Scott, James C. 1998. *Seeing Like a State: How Certain Schemes to Improve the Human Condition have Failed*. In *Yale ISPS Series*. New Haven, CN: Yale University Press.

Searle, John R. 1983. *Intentionality: An Essay in the Philosophy of Mind*. Cambridge: Cambridge University Press.

Singer, P. N. 2022. The Relation Between Perceptual Experience and *Logos*: Galen's Clinical Perspective. In *Galen's Epistemology: Experience, Reason, and Method in Ancient Medicine*, ed. R. J. Hankinson and Matyáš Havrda, 156–189. Cambridge: Cambridge University Press.

Sissons, Jeffrey. 2015. Personhood as History: Māori Conversion in Light of the Polynesian Iconoclasm. *Journal of the Polynesian Society* 124 (2): 129–146.

Spelke, Elizabeth S. 1998. Nativism, Empiricism, and the Origins of Knowledge. *Infant Behavior and Development* 21 (2): 181–200.

Spelke, Elizabeth S., and Katherine D. Kinzler. 2007. Core Knowledge. *Developmental Science* 10 (1): 89–96.

von Staden, Heinrich. 1975. Experiment and Experience in Hellenistic Medicine. *Bulletin of the Institute of Classical Studies* 22:178–199.

Stevens, Stanley S. 1958. Measurement and Man. *Science* 127 (3295): 383–389.

Such, Jan. 2004. *Multiformity of Science*. In *Poznań Studies in the Philosophy of the Sciences and the Humanities*, vol. 79. Amsterdam: Rodopi.

Sutton, Peter, and Keryn Walshe. 2021. *Farmers or Hunter-Gatherers? The Dark Emu Debate*. Melbourne: Melbourne University Press.

Tieleman, Teun. 2002. Galen on the Seat of the Intellect: Anatomical Experiment and Philosophical Tradition. In *Science and Mathematics in Ancient Greek*

Culture, ed. C. J. Tuplin, T. E. Rihll, and Lewis Wolpert, 256–273. Oxford: Oxford University Press.

Treadwell, Jeremy. 2017. Cosmology and Structure: The *Tāhuhu* in the 19th-Century *Whare* Māori. *Journal of the Polynesian Society* 126 (1): 93–122.

Treese, Steven A. 2018. *History and Measurement of the Base and Derived Units.* Cham: Springer.

Tversky, Amos, and Daniel Kahneman. 1982. *Judgment Under Uncertainty: Heuristics and Biases.* In *Judgment Under Uncertainty: Heuristics and Biases*, ed. Daniel Kahneman, Paul Slovic, and Amos Tversky, 3–20. New York: Cambridge University Press.

van der Eijk, Philip J. 2005. *Medicine and Philosophy in Classical Antiquity: Doctors and Philosophers on Nature, Soul, Health and Disease.* Cambridge: Cambridge University Press.

van der Merwe, Ragnar. 2024. Stance Pluralism, Scientology, and the Problem of Relativism. *Foundations of Science* 29 (3): 625–644.

Wiredu, J. E. 1997. How Not to Compare African Traditional Thought with Western Thought. *Transition* 75–76:320–327.

Wootton, David. 2010. Accuracy and Galileo: A Case Study in Quantification and the Scientific Revolution. *The Journal of the Historical Society* 10 (1): 43–55.

Wootton, David. 2015. *The Invention of Science: A New History of the Scientific Revolution.* New York: HarperCollins.

CHAPTER 4

A Very Particular History

Abstract Despite the commonalities between scientific practice and epistemic practices found everywhere, modern science does have distinctively Western features, the products of a very particular history. A first feature of this history was a legal revolution that allowed for the creation of semi-autonomous corporate entities, including universities and scientific societies. A second was a passion for quantification, which emerged from the commercial revolution in late medieval Europe. A third was the growth of philosophical scepticism, which was (in part) a response to challenges to religious authority. As a result of these developments, early modern science was characterized by an emphasis on mechanism, the practice of organized scepticism, and a separation between fact and value, the scope of science being thought of as merely 'matters of fact'.

Keywords Disputation • Exploration • Fact and value • Law • Markets • Mechanism • Quantification • Scepticism

What have we found so far? Advocates of decolonization—whom I have called 'relativizers'—assert that modern science is a Western affair. My argument has been that this is not the case on the operational level of scientific practice (Sect. 2.3.1), where modern science draws on ancient traditions of varied origin, which are refined versions of practices in which all

G. W. Dawes, *Is Science Western?*,
https://doi.org/10.1007/978-3-032-10641-4_4

societies engage. But in this chapter I shall make a concession to the relativizers' case, arguing that science does have some distinctively Western features. These are found not so much on its operational as on its theoretical and metatheoretical levels. Particularly when it comes to its metatheoretical assumptions, modern science continues to bear the marks of its origins in early modern Europe.

4.1 Conditions for the Emergence of Science

Srđan Lelas has argued that modern science must be understood as the product of a particular kind of society. More precisely, it is the product of a particular 'communal mode of living', a way of interacting with one's environment so as to sustain life (Lelas 2000: 176). Lelas suggests that there have been four such modes of living in human history:

> the mode of living of (i) hunter-gatherers of the palaeolithic era, (ii) neolithic farmers and peasants of later times, (iii) aristocrats of agricultural and feudal civilisations, and (iv) entrepreneurs and businessmen of modern time. (Lelas 2000: 177)

In Lelas's view, science in the broad sense of the word —that of the Greek *epistēmē* or the Latin *scientia* (Sect. 1.1.2)—begins to emerge as a result of the urban revolution characteristic of stage (iii). But modern science, as a particular form of *epistēmē*, emerges only as a result of the commercial developments that occur in early modern Europe (at stage [iv]) (Lelas 2000: 180, 194).

I shall return to the role of commerce in the emergence of modern science. But whatever one makes of Lelas's argument, there is an important insight that underlies it. Not all practices are possible, or are equally likely to be found, at all times or in all societies. Take, for example, agriculture. This could exist only in a society that has mastered the art of domestication (Harris 2007: 20). It is also more likely to have arisen in a society of a reasonable size with established property rights (Gallagher et al. 2015: 14222). So what about the practice of modern science? What were the conditions that enabled its emergence? If some of those conditions were distinctively European, they may have given modern science distinctively Western characteristics.

Historians have singled out many factors as conditions for the rise of modern science. A first is that mentioned by Lelas, namely the development of a market economy in late medieval Europe (Lelas 2000: 194). Firstly, this encouraged the standardization of measures (Sect. 3.4.1), made necessary by the development of trade (Ashworth 2004: 1314). Secondly, it led to the development of double-entry bookkeeping, which was an attempt to make 'the real world, the world of bolts of silk, bales of wool and bags of sugar, mathematically legible' (Wootton 2015: 164). Bookkeeping may not appear to have much to do with science, but no less a figure than Galileo appeals to this practice to defend his scientific method (Wootton 2015: 164). Bookkeeping practices are one example of what I shall call 'a passion for quantification' (Sect. 4.2.2), characteristic of this period.

A second factor singled out by historians are the technological developments that occurred in late medieval Europe, developments which went hand-in-hand with the emergence of guilds of skilled craftsmen (Gies and Gies 1994: 124–125). These technological developments made possible the production of scientific tools. Without late medieval work on lenses, for example, the telescope could not have been invented (Gies and Gies 1994: 289). One of these developments, namely the use of printing with moveable type—a technology invented but comparatively little used in China (Chartier 1996: 1)—facilitated the transmission of ideas (Eisenstein 1979: 520). Technological developments in building and warfare encouraged the development of the science of mechanics. These developments were associated with guilds of craftsmen, whose relative autonomy—later to provide a model for universities—was the result of a legal revolution in late medieval Europe (Sect. 4.2.1). But the example of the craftsman also encouraged the 'hands-on' attitude of the experimenter (Smith 2004: 220) and the use of mechanical metaphors for thinking about the natural world (Sect. 4.3.1).

A third condition noted by historians has to do with the early modern European voyages of exploration. In the course of their journeys in pursuit of trade and political power, travellers observed many phenomena—animals, plants, geographical features, forms of social life—that their previous 'experience did not equip them to understand' (Crosby 1997: 51). This not only provided a wealth of new data for scientific inquiry; it helped create the idea that there could exist a continuous programme of coming to know new facts. Contemporaries also noted that these discoveries were made as a result of first-hand experience rather than by consulting ancient texts. As the French-Breton explorer Jacques Cartier wrote in 1545, 'the simple sailors of today have learned the opposite of the opinion of the philosophers by true experience' (Wootton 2015: 81). The authorities on whom the philosophers had relied had also been proved wanting. For the experience of these 'simple sailors' made it 'increasingly clear how much there was of which the Greeks and Romans were either ignorant or misinformed' (Scammell 1969: 393).

A fourth set of conditions mentioned by historians have to do with late medieval and early modern religious thought. I shall discuss one of these—the development of a sceptical spirit—shortly (4.2.3). But there are other features of the religious world that may have paved the way for the scientific revolution. Particularly in England, the Puritan ethos—'a practical utilitarianism devoted to the honor of God' (Cohen 1994: 315)—may have helped motivate the pursuit of 'useful' knowledge (Merton 1938: 431). Late medieval 'voluntarist' theology, with its emphasis on the absolute power of God, may have encouraged empirical investigation, for 'the product of [God's] arbitrary will is contingent, and of the contingent there can be no knowledge beyond experience' (Foster 1936: 5). Empirical investigation may also have been encouraged by the Reformers' revival of the Augustinian doctrine of original sin, with its emphasis on the weakness of human reason (Harrison 2002: 258). Finally, an emphasis on observation and experiment may have been a reaction to the bitter religious disputes of the age, experiment being presented as a way which people of goodwill could agree on 'matters of fact' (Shapin and Schaffer 1985: 22).

4.2 Three European Developments

All of these accounts may be true: they may all identify causal influences that played a role in the rise of modern science (Henry 2009: 101). Historians continue to discuss their relative importance, but I shall not enter into those debates. Rather, I shall zero in on three of these factors, since they do appear to have given early modern science distinctively 'Western' features. The first factor is the legal revolution that made possible the development of quasi-autonomous institutions, such as guilds of craftsmen and universities. The second is the passion for quantification across a wide range of cultural domains. The third is the growth of scepticism, coupled with a critical attitude to authority, which was a product (in part) of the religious crises of the age.

4.2.1 Law

An important feature of modern science is that its practitioners form a specialized class, working in organizations that enable cooperation as well as competition, and whose work has (at least traditionally) eschewed ethical, political, and religious concerns (Sect. 4.3.3). Science has become what one scholar has called 'a self-perpetuating domain of culture', international in its scope (Sect. 1.3) and relatively 'independent ... of the variations in its environment' (Ben-David 1965: 16).

One of the factors that made this possible was the legal revolution that occurred in late medieval Europe. There are two features of that legal revolution that helped create an environment in which modern science could develop. The first was the emergence of a legal science, 'science' here being understood in its broader sense (Sect. 1.1). The initial goal of this science was that of reconciling often conflicting legal sources, a goal summed up in the title of the work written by the twelfth-century scholar Gratian: *Concordia discordantium canonum* (*The Harmony of Discordant Canons*). But it eventually went beyond reconciliation, seeking to understand and systematize the legal principles involved (Kuttner 1941: 16).

This legal science was both the product of, and a contributing factor to, a new 'culture of disputation' (Novikoff 2012: 346). This culture found expression in the method of argumentation known as the scholastic method (since it was that of the medieval schools), which involved the posing of questions and the analysis of arguments. This soon became the most common way of dealing with problems in natural philosophy and theology (Grant 2007: 188). There has been some discussion about whether this scholastic form of argument is European in origin. One argument is that it was already found among Muslim legal scholars (Makdisi 1974: 660). Another is that the Muslims learned this form of argument from the Sarvāstivāda school of Buddhist thought in Central Asia (Beckwith 2012: 62–75). But whatever the origins of the scholastic method, its flourishing in Europe was due to a second product of the late medieval legal revolution, which contributed to the establishment of universities (Makdisi 1974: 660).

The revolution in law had been prompted by struggles between the Papacy and what were increasingly coming to be thought of as 'secular' authorities: a struggle for the independence of the Church from non-ecclesiastical control. The formulation of a body of Church law and of a distinction between ecclesiastical and secular law (each with its own courts) encouraged the emergence of a legal pluralism. This found expression not only in the 'dualism of ecclesiastical and secular jurisdictions', but also in the emergence of 'plural corporate groups within the secular realm, each with its own law' (Berman 1983: 44). As one scholar remarks,

> by the thirteenth century, the principle of *ius proprium* or *iura propria* – namely, the practice of (male) self-organization and self-governance... – extended to almost all forms of known collective activity and undertaking throughout Western Europe. ... Each collective undertaking had its own jurisdiction (*jurisdictio*), authority (*gubernaculum*), and the power to regulate itself (*praeceptio*) through a variety of means. (Sabetti 2004: 74)

Fraternities and guilds were among the most important of these, offering status and a degree of independence to communities of artisans.

The earliest European universities—those of Bologna and Paris—were guilds of this kind, enjoying the right to self-governance. This status did not immediately translate into the modern idea of academic freedom. In 1277, for example, Church authorities tried to to prohibit the teaching of Aristotle's works at the University of Paris. But this attempt ultimately failed. One reason it failed was that the university offered a context in which new ideas could be debated by a scholarly community that enjoyed some degree of autonomy.

Those new ideas included discussions of natural philosophy, stimulated by the translation of Aristotle's *libri naturales* (works dealing with the natural world). The study of natural philosophy was just one branch of philosophy, taught in the arts faculty. As an arts subject, it was (in principle) 'only preparatory to higher studies in theology, law and medicine' (Ben-David 1971: 49). But philosophy increasingly came to be thought of as a subject having principles of its own, distinct from those of theology, whose practitioners formed a distinct community (Ben-David 1971: 50). By the sixteenth century university chairs had been established, particularly in Italy, specializing in subjects such as mathematics, astronomy, and Aristotelian physics (Ben-David 1971: 51). This distinction between natural philosophy and other fields of study was 'a precondition of the emergence of science as an independent intellectual field' (Ben-David 1971: 50).

4.2.2 *Quantification*

We have seen that the use of mathematics in investigating the natural world has ancient roots (Sect. 3.1.1). It is also related to practices of measurement that we find everywhere (Sect. 3.3.2). But the central role of mathematics within modern science arose from a passion for quantification that gripped European society in the fourteenth and fifteenth centuries. By 'quantification' I mean the process of interpreting the multifaceted world of perceptual experience—a world of qualities such as colour, heat, taste, and smell—in such a way that those qualities could be measured numerically.

Today we take such practices for granted. We no longer think of heat and cold, for example, as distinct qualities, for we measure both on a single scale under the heading of 'temperature'. We even describe colours numerically (on the Kelvin scale) for the purpose of lighting and photography. But there is nothing obvious about such a practice. Unlike trees, horses, or apples, which everyday perception reveals as discrete individuals, conditions such as hot and cold, colour, or brightness, are not experienced as entities that can be counted (Crosby 1997: 13). It is evident to everyday perception that such qualities vary in their intensity—one can be more or less hot or cold—and all societies have ways of registering this fact (Sect. 3.3.2). But late medieval thinkers developed a particular interest in attaching numbers and measures to these differing intensities. Indeed some thinkers attempted to find ways of quantifying not only the realities discussed by natural philosophers—such as motion, heat, light, and colour—but also to 'less tractable qualities such as 'love, charity, [and] beauty' (Murdoch 1974: 64).

This process of quantification was facilitated by the importation of the Hindu-Arabic system of numerals, a leading advocate of which was Leonardo Pisano, known as Fibonacci (ca. 1170–1250), although in many domains (such as that of late medieval physics) it involved more geometry than arithmetic. The use of geometry was particularly important for the development of linear perspective in painting, a practice that had ancient roots but which was perfected only during the Renaissance (Edgerton 1975: 64–78). But the passion for quantification was not restricted to the work of academics and painters. It had already become part of everyday life through commercial practices and the measurement of time.

The tenth to the twelfth centuries Northern Europe had seen a commercial revolution (Gies and Gies 1994: 105–108) and it was with the growth of commerce that weighing, numbering, and measuring became particularly important. The relation between quantification and commerce is evident in a dialogue written by Nicholas of Cusa in 1450. When the central character, a humble artisan, wants to instruct a learned humanist, he takes him down to the marketplace, to see 'money being counted ..., goods being weighed; and ... oil and other items being measured out'. The artisan points out that this is how 'men excel the beasts. For brute animals cannot number, weigh, and measure' (Nicholas of Cusa 1996: 488–489 [*Idiota* I 3]).

With regard to time, all humans mark the passage of time by celestial events: sunrise, sunset, solstices, the phases of the moon, and the rising and setting of stars. (My own country has just celebrated the Māori festival of Matariki: the new year marked by the re-appearance of the constellation of the Pleiades.) But in late medieval Europe time also began to be measured arithmetically. After the mechanical clock devised by Henry de Vick had been set up on the Palais Royale in Paris in 1370, 'King Charles V ordered all churches in Paris to ring the hours and quarters according to the time by de Vick's clock' (Crombie 1961: 155).

Although this passion for quantification was found 'in the streets' (as Nicholas's artisan puts it), its relation to the scientific revolution is most evident in the science of mechanics. The growth of commerce stimulated the development of new and improved machines machines: watermills, looms for weaving, and tools for metalwork and glass-blowing. Political instability also stimulated the building of fortifications and the production of weapons of war. All these developments required mechanical know-how.

By the fifteenth century that mechanical know-how was becoming an established science, with its own body of literature. A remarkable body of such mechanical literature had been produced by scribes, even before the proliferation of printing (Long 1997: 2). Its topics

> included gunpowder and artillery; machines, such as mills, pumps, fountains, and cranes; machine parts, particularly gears and pulleys; vehicles, including war carriages, carts, cars, and boats; and numerous other devices. (Long 1997: 2)

This science of machines again had ancient roots (Sect. 3.1.1), and its development was given new status by the translation of the Aristotelian work *Mechanical Problems* (believed at that time to be by Aristotle himself) and the eighth book of Pappus of Alexandria's *Mathematical*

Collection (Van Dyck 2022: 1223). But it became particularly prominent in late medieval and early modern times and would, at a later date, play an important role in the shaping of modern scientific practice (Sect. 3.1). Galileo, for example, frequently mentions Archimedes (287–212 BCE), perhaps the most famous of ancient mechanical writers, praising him for his 'use of mathematics in the treatment of physical problems' (Clagett 1959: 419).

Although there is a connection between the science of mechanics and the 'mechanical philosophy' it later inspired (Sect. 4.3.1), the two are not identical. Firstly, the science of mechanics dealt with artificial contrivances rather than natural phenomena, which means that its scope was more limited than that of the mechanical philosophy. Secondly, the science of mechanics was not initially seen as an alternative to Aristotelian natural philosophy. Mechanics was regarded as one of the Aristotelian 'subalternate' sciences (Sect. 3.1.1), which were distinguished from physics, the study of *phusis* (nature). The two disciplines also tended to be practised by different people: mathematicians on the one hand, and natural philosophers on the other (Gabbey 1993: 134).

This disciplinary divide was only overcome in the seventeenth century. Galileo's insistence that he be given the title of 'mathematician and philosopher' by the Grand Duke of Tuscany was a sign of the merger of what had previously been distinct occupations. But even before Galileo's time, some sixteenth-century writers on mechanics had begun to display an interest in questions that bordered on natural philosophy. One German writer on weaponry, for example, asked whether it is 'the fire or the force of the vapor produced by the fire that drives the stone from the cannon' and proposed an experiment to settle the question (Long 1997: 18). In asking such questions, writers on mechanics prepared the way for a natural philosophy inspired by mechanical principles (Sect. 4.3.1).

4.2.3 Scepticism

The fifteenth century saw the birth—or, more accurately, the rebirth—of a tradition of philosophical scepticism, arguing for limits on what we can know. Richard Popkin traces the origins of early modern scepticism to both the Protestant Reformation of the sixteenth century and the rediscovery of the tradition of philosophical scepticism found particularly in the work of Sextus Empiricus (ca. 150–225 CE). But as Popkin himself notes (2003: xix), the sceptical spirit predates both of these. Its roots can be found in late medieval developments in theology and in the religious crisis known as the Western Schism.

The fourteenth century saw a movement away from the synthesis of faith and reason—in particular of Catholic faith and Aristotelian reason—that had been characteristic of the work of Thomas Aquinas. Influenced by an increasing theological suspicion of Aristotelian doctrines (Swieżawski 2022: 8), fourteenth-century thinkers developed a theology shaped by criticism and scepticism (Leff 1956: 32). The scepticism was directed in the first place towards our any claimed ability to know the ways of God. An uncompromising insistence on God's absolute power led thinkers to suggest that we cannot predict his actions (Leff 1956: 37–38). They can be known only by observing what actually occurs. But this scepticism also took the form of a challenge to the assumed certainty of scientific knowledge. The 'nominalist' insistence that only individual entities exist meant that 'all our knowledge of the world must be founded on singular propositions' (Milton 1980: 138). This set the scene for the problem of induction made famous by David Hume (1711–1776) (Milton 1980: 141): how can we know that our general statements about the world—what would later be known as 'laws of nature'—are true?

These trends in late medieval thought were reinforced by the fifteenth-century rediscovery of the tradition of ancient philosophical scepticism. Ancient Stoic philosophers had argued that there was a certain kind of sense impression that carried with it an assurance of truth and could serve as a firm foundation for knowledge (Sextus 1933: 133 [VII 1.248]). When I see an apple, for instance, I immediately perceive it as an apple and a vivid impression of this kind is reliable. Those known as 'Academic' sceptics denied this, arguing that there was no way of distinguishing apparently certain sense impressions from those that were deceptive: for every apparently reliable impression we could think of a fake one indistinguishable from it (Sextus 1933: 134–137 [VII 1.252]). Another group of sceptics—known as 'Pyrrhonian sceptics'—went further, arguing that even the claim that there is no reliable criterion of knowledge is a form of 'dogmatism' (Sextus 1976: 3; *PH* 1.3), since it is itself a claim to knowledge. Pyrrhonian sceptics employed forms of reasoning intended to show that neither side of any argument could make its case in a non-question-begging manner. The appropriate response, they argued, is a suspension of judgement (*epochē*). Once you suspend judgement, you will find that you have achieved a state of calmness of mind: 'unperturbedness' or 'quietude' (*ataraxia*).

The rediscovery of this tradition of ancient scepticism and its popularization, particularly through the work of Michel de Montaigne (1533–1592), brought about what Popkin calls *la crise pyrrhonienne* of the seventeenth century' (Popkin 2003: 43). Even in its more moderate forms, this new sceptical spirit drew attention to the limits of reason. The combined effects of late medieval nominalism and moderate scepticism can be found in the work of John Locke (1632–1704), whose distinction between 'nominal' and 'real' essences—scientific theories being restricted to the former—merely made explicit what William of Ockham (1285–1347) and his followers had said (Milton 1980: 139). This sceptical spirit was not immediately corrosive of religious belief. Indeed early modern sceptics frequently argued that knowing the limits of human reason should make us more dependent on religious faith (Popkin 2003: xxi). But by casting doubt on the powers of reason, the sceptics widened the gap between faith and reason that had already been opened up in the fourteenth century.

The problem with appeals to faith is that confidence in traditional forms of religious authority had also been eroded. It was eroded, first of all, by the Western Schism: the period between 1378 and 1417 when there were at first two and later three individuals claiming to be the duly elected Pope. For Catholics anxious to do the right thing, it was far from clear how to proceed, for at least two of the claimants 'had arguments and competent lawyers on their side' (Schüssler 2009: 449). The theological crisis this provoked led to the development of a doctrine of 'moral certainty' by the theologian Jean Gerson (1363–1429). This held that individuals could adopt in good conscience 'the opinion they regarded as the most reasonable, as long as it was supported by a reputable group of experts' (Schüssler 2009: 452). In the course of defending this level of certainty, Gerson made reference to the Academic sceptics, recognizing that the question he faced resembled theirs (Schüssler 2009: 459).

The erosion of confidence in Papal authority precipitated by these events was reflected in the decrees of the Council of Constance, which brought the Schism to an end. These included the claim (supported by Jean Gerson) that a General Council of the Church has an authority even greater than that of a Pope, when it comes to dealing with such a crisis. This challenge to religious authority was exacerbated by the Protestant Reformation, which was itself a 'dispute over the proper standard of religious knowledge, or what was called "the rule of faith"' (Popkin 2003: 3). In 1517, Martin Luther offered a new criterion of religious truth, namely that 'what conscience was compelled to believe on reading Scripture is true' (Popkin 2003: 5). But this criterion itself required justification, as Luther's opponents were quick to point out. The refusal of the Reformers to offer a rational defence—their insistence that the authority of Scripture was self-authenticating (Calvin 1960: 80–81 [VII 5])—further widened the gap between faith and reason.

Yet the Reformation, for all its dogmatism, also offered a model of how authority could be challenged. It is no accident that the pioneers of modern science used its language and appealed to its example. As David Wootton writes,

> before the Scientific Revolution was a revolution, it was a reformation. 'The main Design,' wrote [Robert] Hooke in 1665, of his own efforts and those of the Royal Society, was 'a reformation in Philosophy'. Thomas Sprat, writing a history of the Royal Society in 1667, repeatedly compared the reformation in natural philosophy with the earlier reformation in religion. (Wootton 2015: 34)

As we shall see (Sect. 4.3.2), the institutions of modern science facilitate this challenge to traditional authorities.

4.3 Three Western Features

These developments—the emergence of partially autonomous institutions of learning, the quantification of qualities, and the questioning of authority—had some parallels outside of Europe. But as in the case of printing with moveable type (Sect. 4.1), they had a wider and more profound impact in Europe than elsewhere.

Other societies, for example, had institutions of higher education. The Muslim world, for example, had *madāris* (the plural of *madrasa*), while the Buddhist world had *vihārā* (the plural of *vihāra*), which in some places evolved from monasteries to general centres of learning (Sarao 2017: 1329–1330). There were also 'analogues to universities in China, though not many' (Elvin 2004: xxvii). The *vihāra* in China, for example, remained a largely religious institution (Beckwith 2012: 139). But the corporate legal status of European universities set them apart (Dmitrishin 2013: 12), offering a degree of autonomy and a form of 'existence independent of its individual members' (Makdisi 1970: 264). With regard to quantification, Chinese practice had long predated the European, at least in political affairs (Sect. 3.4.1). But 'the drive to what might be called the "mathematization of observations" was significantly weaker in premodern China' than in Europe (Elvin 2004: xxxii). In a similar way, the spirit of criticism—the practice of 'questioning all human activity' (Jaspers 1956: 6)—was common to the three post Axial-Age civilizations of Greece, India, and China. But nowhere outside of Europe did the questioning of authority have such dramatic social and political consequences.

These three historical developments—a legal revolution, a passion for quantification, and a revival of scepticism—gave early modern science three distinctively Western characteristics. The first was a preference for the root metaphor of the machine, which was associated with the 'mathematization' of science. The second was what I shall call (following Robert Merton) the practice of 'organized scepticism'. A third was a separation of fact and value, in the sense that scientific inquiry is supposed to be directed only to the former.

4.3.1 The Mechanical Philosophy

I have argued that the metatheoretical level of scientific practice consists of presuppositions that Gerald Holton refers to as *themata* (Sect. 2.3.1). Holton distinguishes *themata* from metaphors (Holton 1996: 463), metaphor being a broader category. But *themata* can take the form of 'root metaphors' (Pepper 1982, 199): metaphors that shape a domain of inquiry. Holton himself mentions one such metaphor when discussing the work of Johannes Kepler (1571–1630). Kepler, on Holton's account, worked with 'three basic thematic ideas: the universe as a physical machine, the universe as a mathematical harmony, and the universe as central theological order' (Holton 1996, 457). At least the first of these is a root metaphor.

Human thought about the natural world has been shaped by a surprisingly small number of root metaphors. Perhaps the most ancient of these, widely used in mythical cosmologies, is 'that of organic growth and reproduction', often sexual reproduction (Hesse 1962: 31). Mary Hesse refers to this as 'the analogy of organism', which passes out of mythology to play an important role in the philosophy of Aristotle (Marques and Brito 2014: 78). A second type of root metaphor is what Hesse calls 'the analogy of attraction'.

> Men experience sympathy and antipathy, attraction and repulsion, love and strife, between themselves and other men, and between themselves and

> nature, and these are therefore seen as forces which can produce effects in nature. (Hesse 1962: 31)

This metaphor is also very ancient, being one of the principles of magic, which is often thought to operate by way of 'sympathies' and 'antipathies' between elements of the natural world. The same metaphor found expression in the work of the Presocratic philosopher Empedocles (ca. 495–435 BCE), for whom the two basic principles shaping reality are 'love' and 'strife'. A third root metaphor is that of the spoken word, by which a ruler, in particular, can make things happen merely by speaking. In the first chapter of Genesis, for example, God creates the world by way of the spoken word. Finally, there are root metaphors that arise from 'the techniques and artefacts of the human workman' (Hesse 1962: 33). Again, these can be found in myth: in Genesis 2, God forms Adam from clay, in the manner of a potter. But metaphors derived from human artefacts also form the basis of mechanical explanations. As Descartes wrote, 'in artifacts made by craftsmen the machines are large and visible, whereas in nature they are small and often invisible' (Bertoloni Meli 2006: 138). But in both cases effects are brought about by 'contact among bodies' (Bertoloni Meli 2006: 135).

Historians writing about the scientific revolution have often depicted it as a struggle between two root metaphors: an Aristotelian metaphor of organism and a modern metaphor of mechanism. In support of this idea they could cite Kepler's words: 'my aim … is to show that the celestial machine is to be likened not to a divine organism but rather to a clockwork' (Holton 1988: 56). They could also cite Descartes's remark that 'I have described this earth and indeed the whole visible universe as if it were a machine' (Garber 2013: 16).

One can argue that so simple a contrast does not do justice to the complexity of the intellectual world from which modern science emerged (Garber 2013: 3). Firstly, some pioneers of modern science remained at the 'operational' level of phenomena, without committing themselves to any particular view of underlying causes, whether mechanical or organic (Sect. 2.3.3). Secondly, many who did seek to understand underlying causes were not convinced that purely mechanical explanations would suffice (Henry 1986: 337, 229). Thirdly, there were differences of opinion about what it meant for a science to be 'mechanistic' (Garber 2013: 18–20). But despite these qualifications, one can still argue that the pioneers of early modern science had a preference for explanations modelled on machines (Garber 2013: 3). These thinkers avoided (as far as they could) metaphors derived from living organisms and were suspicious of ideas of sympathy and antipathy (which involved action at a distance). Their preference was for the root metaphor of the machine.

While this preference may have been encouraged by the development of new technology—particularly that of mechanical clocks—the metaphor itself was not new. As we have seen, the work of the craftsman constitutes an ancient root metaphor for thinking about the natural world, a metaphor that mechanical explanations develop. But among ancient thinkers, this was just one metaphor among many (Hesse 1962: 33). While Aristotle, for example, employed mechanical explanations, he regarded these as complementary to teleological explanations based on the metaphor of organism (Johnson 2017: 132–133). What we see in the seventeenth century is a narrowing of the range of range of the root metaphors employed, with a preference for that of mechanism (Hesse 1962: 73).

Why was this? There seem to have been a number of contributing factors. Firstly, the religious context of early modern Europe had a part to play. Piyo Rattansi, for example, has argued that some pioneers of modern science—Robert Boyle (1627–1691) being a striking example—were initially attracted by the worldview of the Hermetic philosophy, with its talk of sympathies and antipathies (Rattansi 1968: 132, 135). But they were later made anxious by the association of this Hermetic philosophy with radical sectarian movements (Sect. 4.3.3). This political dimension to the Hermetic philosophy led them to favour the mechanical philosophy, as 'a sober and disenchanted system of natural knowledge' which could be 'harmonized with traditional religion' (Rattansi 1968: 137).

The idea that the machine metaphor was compatible with traditional religion in particularly evident in Boyle's writings. He complains, for example, about the Aristotelian custom of regarding 'nature' as if she had purposes of her own (Boyle 1996: 52 [§4]), a practice that could lend support to atheism. For Boyle the wisdom of God is more clearly exhibited in his ability to 'make so vast a machine [the universe] perform [so] many things … by the mere contrivance of brute matter' (Boyle 1996: 11 [§1]).

Yet there was a second important factor leading to the dominance of mechanistic explanations, which arose from the passion for quantification characteristic of the late medieval world. Mechanistic explanations were modelled on those offered in the science of mechanics (Sect. 3.1.2). One reason for the popularity of the metaphor was that mechanics was a mathematical science (Johnson 2017: 134). For consistent mechanists, the behaviour of bodies can be reduced to the laws of geometry: 'bodies are the objects of geometry made real' (Garber 2013: 19). The geometry that the artists had used to represent the world (Sect. 4.2.2) could now be used to explain it.

Combined with the new passion for experimentation (Sect. 2.3.3), mechanical explanations also offered the hope that precise predictions could be used to settle scientific disputes (Van Dyck 2022: 1231). The 'analogies of organism and attraction', for example,

> were said to be anthropomorphic and condemned, not really because of their anthropomorphism, but because comparisons with human and animal life could not be given clear and distinct expression and could not be dealt with in mathematical terms, and therefore could not be compared in detail with the facts. (Hesse 1962: 99)

Once again, more than one factor was at play, and the machine metaphor was not the only one employed by seventeenth-century scientists. But it does seem to have been favoured because of its potential for precise mathematical expression.

Was there something distinctively 'Western' about this favouring of the machine metaphor? The passion for quantification was a development of the practice of counting and measurement that all peoples share (Sect. 3.3.2). But it became a passion because of the particular historical circumstances of late medieval Europe (Sect. 4.2.2). More importantly, it seems to have been only in Europe that mathematical practices were coupled with the root metaphor of the machine. Take, for example, China, a civilization whose technological achievements were (for many centuries) ahead of those found in Europe. The Chinese had long possessed precise practices of measurement (Bodde 1991: 139). As early as the third century BCE they had made efforts to standardize weights and measures (Bodde 1991: 137) Chinese thinkers did use analogies drawn from the work of craftsmen to illustrate their philosophy. But for the most part they did so while developing ethical and political theories (Bodde 1991: 136). Although familiar with machines and skilled in the practice of quantification, the Chinese never adopted the machine as a metaphor for thinking about the natural world (Major 1993: 31). The preferred Chinese metaphor for thinking about natural phenomena is that of *gănyìng* ('resonance'), based on the 'sympathetic vibration' of strings on musical instruments (Le Blanc 1995: 59). (It can be a useful metaphor: the psychologist James Gibson, for example, used it in relation to perception [Gibson 1966: 271].) 'Resonance' belongs to the family of root metaphors that Hesse calls the 'analogy of attraction' (Hesse 1962: 31).

4.3.2 Organized Scepticism

Late medieval and early modern Europe saw the rise of a sceptical spirit in both philosophy and religion (Sect. 4.2.3). But its effects would have been transitory had it not been embodied in institutions.

I have already discussed the development of universities, whose corporate structure gave them (and their members) a degree of autonomy vis-à-vis the surrounding society (Sect. 4.2.1). It has been customary in discussions of the scientific revolution to downplay the role of the universities. This may be because attention has been focused on universities of Oxford and Paris or it may be due to the anti-Aristotelian polemics of early modern writers such as Francis Bacon. David Wootton, for example, continues to describe the Aristotelianism of the early modern universities as hostile to new forms of knowledge (Wootton 2015: 71). But recent work has suggested that the universities often provided support to early modern scientists (Feingold and Navarro-Brotóns 2006: 1). In some cases, they were the very places in which these developments occurred. Significant advances in mathematics, for example, were made at the University of Bologna in the sixteenth century (Borgato 2006: 128–129), while some of Galileo's most important discoveries were made while teaching at the University of Padua (Borgato 2006: 130), a period he described as 'the best eighteen years' of his life (Galilei 1906: 209).

It remains the case, however, that much of the new scientific work in early modern Europe took place outside the universities, in newly established academies and scientific societies. Noteworthy among these new organisations was the *Accademia dei Lincei* (Academy of the Lynxes), to which Galileo belonged and which published two of his early works (Ornstein 1928: 75–76), and the closely related *Accademia del Cimento* (Experimental Academy), founded by two of Galileo's students. In France there was the Académie des Sciences, while England had the Royal Society of London. These organizations were descended from the humanist academies—originally informal gatherings of scholars—that had formed around thinkers such as Marsilio Ficino (1433–1499) (Ben-David 1971: 59–60).

These academies and societies offered another venue in which scholars could enjoy a degree of autonomy, outside the corporate structure of the universities, an autonomy guaranteed by the protection of 'princes, great noblemen, or even municipalities' (Ben-David 1971: 60). The new academies quickly became formal institutions (Ben-David 1971: 61). Like the medieval guilds on which the universities were modelled (Sect. 4.2.1), they 'increasingly insisted upon their exclusive prerogatives to judge the work of members and colleagues' and to decide who should count as a member or colleague (Daston 1994: 113).

Despite drawing clear lines between insiders and outsiders, these new societies saw themselves as engaging in what Robert Merton called the practice of 'organized scepticism' (Merton 1942: 126). The goal of the *Accademia dei Lincei* is summed up by the response of its founder, Federico Cesi (1585–1630), to a letter defending Copernican astronomy. 'The author assumes,' Cesi protests, 'that all our companions are Copernicans, despite the fact that's not true. All that we are committed to as a group is freedom in natural philosophy' (Biagoli 1993: 80). In a similar way, the Royal Society held to an ideal of examining all questions first-hand and not being bound by traditional views. Their motto was (and remains) *nullius in verba* ('take no one's word for it'), an allusion to a phrase by the Roman poet Horace (65–68 BC), which was understood as a declaration of independence from authority (Sutton 1994: 57). Authority in the form of testimony did play an important role in the Society's work, for that work relied heavily on reports of experiments (Dear 2015: 63–65). But the authority of witnesses was not a final court of appeal. The ideal was that qualified readers of the testimony could verify the results for themselves (Dear 2015: 64).

The organized scepticism of early modern science was a mitigated scepticism: one that supported modest claims to knowledge by reference to experimental 'matters of fact'. But the practice of scepticism—'the suspension of judgment until "the facts are at hand" and the detached scrutiny of beliefs in terms of empirical and logical criteria' (Merton 1942: 126)—remains a scientific ideal. Nor it is a mere ideal; it is embodied in collective practices (Sect. 3.4.3). As David Hull writes,

> the objectivity that matters so much in science is not primarily a characteristic of individual scientists, but of scientific communities. Scientists rarely refute their own pet hypotheses, especially after they have appeared in print,

> but that is all right. Their fellow scientists will be happy to expose these hypotheses to severe testing. (Hull 1988: 4)

Scientific communities certainly have their own traditions. These includes bodies of what seem (at a particular time) to be well-established theories, which constitute a kind of scientific orthodoxy. Ideas that appear radically incompatible with existing theories will not (and arguably should not) readily be accepted (Polanyi 1963: 1012–1013). But one of the methodological *themata* of modern science is that, in the long run, no theory is beyond criticism. The fact that there have occurred scientific revolutions—radical revisions of conceptual frameworks—shows that this assumption does shape scientific practice.

4.3.3 *The Separation of Fact and Value*

Today's sciences have moved beyond the mechanical philosophy that inspired early seventeenth-century thinkers (Clayton 2013: 441–442). Already in Newton's work, the idea of the machine had become the more abstract idea of a 'system', whose parts 'were not necessarily connected through material links' (Van Dyck 2022: 1230). There remain sciences in which the machine metaphor continues to function, a striking example being biology, where an analogy with machines is often used to explain the workings of the cell (Nicholson 2019: 108–109). Here, too, however, machine metaphor is being supplemented, if not replaced (Clayton 2013: 441). Some philosophers continue to insist that what they call 'mechanistic' explanation is characteristic of modern science (Glennan et al. 2022: 145, 148). But in doing so they employ a very minimal sense of the term 'mechanistic', which no longer involves simple mechanical analogies (Levy 2014: 8–9).

There is, however, another *thema* that continues to shape the practice of science. Like the practice of organized scepticism, it is a methodological *thema*, in the sense of an assumption about the nature of scientific inquiry. It has to do with what one author calls 'the normative (social, political, religious) neutralization of the knowledge of nature' (van den Daele 1977: 28). This entails a sharp distinction between matters of fact and matters of value, the scope of science being restricted to the former. More precisely, it entails a sharp distinction between the kinds of facts that can be discovered by observation and experimentation and the kinds of facts (if there are any) that lie behind judgements of value.

This is a different idea from the claim that the practice of science is itself 'value-free', an idea that has been rightly contested. The practice of science clearly involves epistemic values, of the kind highlighted by Robert Merton (Merton 1942: 118–126). Helen Longino refers to these as 'constitutive values', since they are intrinsic to scientific practice (Longino 1990: 4). But one can argue that the practice of science can also involve non-epistemic values, of the kind Longino calls 'contextual' (Longino 1990: 4). Contextual values can influence the ways scientists identify their object of study (Ludwig 2016: 1266); they can shape the way they choose conceptual frameworks, and acquire data; they can even enter into the very concepts scientific theories employ (De Melo-Martín and Intemann 2016: 514–515). All of these matters, however, lie outside the scope of my discussion, for the point I am making is a different one. It is that the question of what constitutes value—the questions of what is good—is normally thought to lie outside the scope of science. Scientists have traditionally asked about what is, not about what ought to be, regarding the latter as outside of their domain of inquiry (Campbell 1921: 164).

The late medieval thinkers who began the process of quantifying physics (Sect. 4.2.2) applied their methods to a very broad range of questions. They wrote about physics, but not merely about physics. The mathematical optics of Robert Grosseteste (ca. 1165–1253) grew out of his metaphysics of light and theologians applied quantified concepts to topics such as divine grace and sin (Crombie 1961: 147, 154–155). Such inquiries included both matters of fact and questions of value. The Aristotelian natural philosophy that became so influential in the thirteenth century was also shot through with considerations of value. It saw the world as a hierarchy of beings, each of which is moved by what is good, the ultimate good being the unmoved mover (Lloyd 1968: 295).

The idea that a *scientia* should eschew questions about value appears to have emerged only in the Europe, being foreign to systems of knowledge found elsewhere (Harding 1997: 48). Chinese Confucians, for example, investigated the principles of nature to discover how we should live (Son 1986: 154). But this separation of fact and value was also foreign to the European intellectual movements that immediately preceded modern science. Practitioners of the Hermetic philosophy, for example, sought not merely knowledge of nature, but also the transformation of society (van den Daele 1977: 38). It was for precisely this reason that Boyle and his colleagues were worried about it (Sect. 4.3.1).

So how did early modern science come to distinguish between matters of fact and matters of value? How did it come to regard itself as concerned only with the former? One can argue that this third *thema* was related to the first I mentioned: the growth of the mechanical philosophy was one of the factors encouraging scientists to set aside questions of value. Living organisms exhibit at least rudimentary forms of agency, seeking what seems good for them (Kauffman 2013: 6). But mere matter in motion seeks no goals (Proctor 1991: 7, 41). It follows that appeal to values (to what seems good) has no explanatory role within a mechanistic science. This role is now gradually being recovered (Kauffman 2000: 118), with the rise of systems theory. But it is being recovered because systems theory began by replacing the machine metaphor with that of the organism (Bausch 2002: 418).

So the machine metaphor itself may have played a role in banishing values from scientific inquiry. But here, too, social and political factors appear to have been important. With regard to England, the virtuosi of what became the Royal Society sought to gain protection and have a secure place in post-Restoration society. But to do so they had to abandon the idea that their new science would contribute to religious and social reform. Whatever religious and social ideals may once have motivated them now needed to be set aside for the sake of political stability. So the pioneers of the new science deliberately left ethical and religious questions outside the scope of their inquiries (Van den Daele 1977: 41).

In drafting statutes for the Royal Society, for example, Robert Hooke noted that it would avoid 'meddling with Divinity, Metaphysics, Moralls, Politicks, Grammar, Rhetoric, or Logick' (Proctor 1991: 33). A similar restriction of the scope of science occurred in continental Europe. The Accademia del Cimento claimed it would avoid 'Controversie with any, or engage in any Nice Disputation, or heat of Contradiction', while Christiaan Huygens (1629–1695) suggested that the Académie des Sciences avoid discussing 'the secrets of religion' or 'affairs of the state' (Proctor 1991: 34).

These restrictions were partly related to an emerging ideal of the character of the scientist. 'The calm of the natural philosopher' stood in contrast to 'the reformist zeal of the political or religious enthusiast' (Proctor 1991: 33). Here, too, the religious controversies of the age played a role. A society less racked by religious strife would have been less likely to produce such an ideal.

References

Ashworth, William J. 2004. Metrology and the State: Science, Revenue, and Commerce. *Science* 306 (5700): 1314–1317.

Bausch, Kenneth C. 2002. Roots and Branches: A Brief, Picaresque, Personal History of Systems Theory. *Systems Research and Behavioral Science* 19 (5): 417–428.

Beckwith, Christopher I. 2012. *Warriors of the Cloisters: The Central Asian Origins of Science in the Medieval World*. Princeton, NJ: Princeton University Press.

Ben-David, Joseph. 1965. The Scientific Role: The Conditions of its Establishment in Europe. *Minerva* 4 (1): 15–54.

Ben-David, Joseph. 1971. *The Scientist's Role in Society: A Comparative Study. Foundations of Modern Sociology*. Englewood-Cliffs, NJ: Prentice-Hall.

Berman, Harold J. 1983. *Law and Revolution: The Formation of the Western Legal Tradition*. Cambridge, MA: Harvard University Press.

Bertoloni Meli, Domenico. 2006. *Thinking with Objects: The Transformation of Mechanics in the Seventeenth Century*. Baltimore, MD: Johns Hopkins University Press.

Biagoli, Mario. 1993. *Galileo, Courtier: The Practice of Science in the Culture of Absolutism. Science and Its Conceptual Foundations*. Chicago: University of Chicago Press.

Bodde, Derk. 1991. *Chinese Thought, Society, and Science: The Intellectual and Social Background of Science and Technology in Pre-modern China*. Honolulu, HI: University of Hawaii Press.

Borgato, Maria Teresa. 2006. Mathematical Research in Italian Universities in the Modern Era. In *Universities and Science in the Early Modern Period*, Archimedes: New Studies in the History and Philosophy of Science and Technology, ed. Mordechai Feingold and Victor Navarro-Brotóns, vol. 12, 127–139. Dordrecht: Springer.

Boyle, Robert. 1996. *A Free Enquiry into the Vulgarly Received Notion of Nature*. In *Cambridge Texts in the History of Philosophy*, ed. Edward B. Davis and Michael Hunter. Cambridge: Cambridge University Press.

Calvin, John. 1960. *Institutes of the Christian Religion*. Vol. 1. Translated by Ford Lewis Battles. Library of Christian Classics 20. Philadelphia, PA: The Westminster Press.

Campbell, Norman. 1921. *What is Science?* London: Methuen & Co.

Chartier, Roger. 1996. Gutenberg Revisited from the East. Translated by Jill A. Friedman. *Late Imperial China* 17(1): 1–9.

Clagett, Marshall. 1959. The Impact of Archimedes on Medieval Science. *Isis* 50 (4): 419–429.

Clayton, Philip. 2013. Machines and Organisms: The Rise and Fall of a Conflict. In *Beyond Mechanism: Putting Life Back into Biology*, ed. Brian G. Henning and Adam C. Scarfe, 434–447. Lanham, MD: Lexington Books.

Cohen, H. Floris. 1994. *The Scientific Revolution: A Historiographical Inquiry*. Chicago: University of Chicago Press.

Crombie, Alistair C. 1961. Quantification in Medieval Physics. *Isis* 52 (2): 143–160.

Crosby, Alfred W. 1997. *The Measure of Reality: Quantification and Western Society, 1250–1600*. Cambridge: Cambridge University Press.

Daston, Lorraine. 1994. The Several Contexts of the Scientific Revolution. *Minerva* 32 (1): 108–114.

De Melo-Martín, Inmaculada, and Kristen Intemann. 2016. The Risk of Using Inductive Risk to Challenge the Value-Free Ideal. *Philosophy of Science* 83 (4): 500–520.

Dear, Peter. 2015. *Totius in verba*: Rhetoric and Authority in the Early Royal Society. In *Rhetoric and the Early Royal Society: A Sourcebook*, Scholarly Communication, ed. Tina Skouen and Ryan Stark, vol. 3, 53–76. Leiden: E. J. Brill.

Dmitrishin, Alexander. 2013. Deconstructing Distinctions: The European University in Comparative Historical Perspective. *Entremons: UPF Journal of World History* 5:1–18.

Edgerton, Samuel Y. 1975. *The Renaissance Rediscovery of Linear Perspective*. New York: Basic Books.

van der Eijk, Philip J. 2005. *Medicine and Philosophy in Classical Antiquity: Doctors and Philosophers on Nature, Soul, Health and Disease*. Cambridge: Cambridge University Press.

Eisenstein, Elizabeth L. 1979. *The Printing Press as an Agent of Change: Communications and Cultural Transformations in Early-Modern Europe*. Cambridge: Cambridge University Press.

Elvin, Mark. 2004. Vale atque Ave. In *Science and Civilization in China, by Joseph Needham*, ed. Kenneth Girdwood Robinson, vol. 7, part 2, xxiv–xliii. Cambridge: Cambridge University Press.

Feingold, Mordechai, and Victor Navarro-Brotóns. 2006. Introduction. In *Universities and Science in the Early Modern Period*, Archimedes: New Studies in the History and Philosophy of Science and Technology, ed. Mordechai Feingold and Victor Navarro-Brotóns, vol. 12, 1–7. Dordrecht: Springer.

Foster, M. B. 1936. Christian Theology and Modern Science of Nature (II). *Mind* 45 (177): 1–27.

Gabbey, Alan. 1993. Between *ars* and *philosophia naturalis*: Reflections on the Historiography of Early Modern Mechanics. In *Renaissance and Revolution: Humanists, Scholars, Craftsmen and Natural Philosophers in Early Modern Europe*, ed. J. V. Field and Frank A. J. L. James, 133–145. Cambridge: Cambridge University Press.

Galilei, Galileo. 1906. Galileo to Fortunio Liceti, 23 June 1640. In *Le Opere de Galileo Galilei: Edizione Nazionale*, ed. Antonio Favaro, vol. 18, 207–209. Firenze: Tipografia Barbèra.

Gallagher, Elizabeth M., Stephen J. Shennan, and Mark G. Thomas. 2015. Transition to Farming more Likely for Small, Conservative Groups with Property Rights, but Increased Productivity is not Essential. *Proceedings of the National Academy of Sciences* 112 (46): 14218–14223.

Garber, Daniel. 2013. Remarks on the Pre-history of the Mechanical Philosophy. In *The Mechanization of Natural Philosophy*, Boston Studies in the Philosophy of Science, ed. Daniel Garber and Sophie Roux, vol. 300, 3–26. Dordrecht: Springer.

Gibson, James J. 1966. *The Senses Considered as Perceptual Systems.* London: George Allen & Unwin.

Gies, Frances, and Joseph Gies. 1994. *Cathedral, Forge, and Waterwheel: Technology and Invention in the Middle Ages.* New York: HarperCollins.

Glennan, Stuart, Phyllis Illari, and Erik Weber. 2022. Six Theses on Mechanisms and Mechanistic Science. *Journal for General Philosophy of Science* 53:1–19.

Grant, Edward. 2007. *A History of Natural Philosophy: From the Ancient World to the Nineteenth Century.* Cambridge: Cambridge University Press.

Harding, Sandra. 1997. Is Modern Science an Ethno-Science? Rethinking Epistemological Assumptions. In *Science and Technology in a Developing World*, Sociology of the Sciences, ed. Terri Shinn, Jack Spaapen, and Venni Krishna, vol. 19, 37–64. Dordrecht: Springer.

Harris, David R. 2007. Agriculture, Cultivation and Domestication: Exploring the Conceptual Framework of Early Food Production. In *Rethinking Agriculture: Archaeological and Ethnoarchaeological Perspectives*, ed. Tim Denham, José Iriarte, and Luc Vrydaghs, 16–35. London: Routledge.

Harrison, Peter. 2002. Original Sin and the Problem of Knowledge in Early Modern Europe. *Journal of the History of Ideas* 63 (2): 239–259.

Henry, John. 1986. Occult Qualities and the Experimental Philosophy: Active Principles in pre-Newtonian Matter Theory. *History of Science* 24 (4): 335–381.

Henry, John. 2009. Voluntarist Theology at the Origins of Modern Science: A Response to Peter Harrison. *History of Science* 47 (1): 79–113.

Hesse, Mary B. 1962. *Forces and Fields: The Concept of Action at a Distance in the History of Physics.* Mineola, NY: Dover Publications.

Holton, Gerald. 1988. *Thematic Origins of Scientific Thought: Kepler to Einstein.* Revised ed. Cambridge, MA: Harvard University Press.

Holton, Gerald. 1996. The Role of Themata in Science. *Foundations of Physics* 26 (4): 453–465.

Hull, David L. 1988. *Science as a Process: An Evolutionary Account of the Social and Conceptual Development of Science. Science and Its Conceptual Foundations.* Chicago: University of Chicago Press.

Jaspers, Karl. 1953. *The Origin and Goal of History.* Translated by Michael Bullock. New Haven, CN: Yale University Press.

Johnson, Monte Ransome. 2017. Aristotelian Mechanistic Explanation. In *Teleology in the Ancient World: Philosophical and Medical Approaches*, ed. Julius Rocca, 125–150. Cambridge: Cambridge University Press.

Kauffman, Stuart A. 2000. *Investigation*. Oxford: Oxford University Press.

Kauffman, Stuart A. 2013. Evolution beyond Newton, Darwin, and Entailing Law. In *Beyond Mechanism: Putting Life Back into Biology*, ed. Brian G. Henning and Adam C. Scarfe, 1–24. Lanham, MD: Lexington Books.

Kuttner, Stephan. 1941. The Father of the Science of Canon Law. *Jurist* 1 (1): 2–19.

Le Blanc, Charles. 1995. From Cosmology to Ontology through Resonance: A Chinese Interpretation of Reality. In *Beyond Textuality: Asceticism and Violence in Anthropological Interpretation*, Approaches to Semiotics, ed. Gilles Bibeau and Ellen Corin, vol. 120, 57–77. Berlin: Mouton de Gruyter.

Leff, Gordon. 1956. The Fourteenth Century and the Decline of Scholasticism. *Past & Present* 9:30–41.

Lelas, Srđan. 2000. *Science and Modernity: Toward an Integral Theory of Science*. In *Boston Studies in the Philosophy of Science*, vol. 214. Dordrecht: Kluwer.

Levy, Arnon. 2014. Machine-Likeness and Explanation by Decomposition. *Philosopher's Imprint* 14 (6): 1–15.

Lloyd, G. E. R. 1968. *Aristotle: The Growth and Structure of His Thought*. Cambridge: Cambridge University Press.

Long, Pamela O. 1997. Power, Patronage, and the Authorship of *Ars*: From Mechanical Know-How to Mechanical Knowledge in the Last Scribal Age. *Isis* 88 (1): 1–41.

Longino, Helen E. 1990. *Science as Social Knowledge: Values and Objectivity in Scientific Inquiry*. Princeton, NJ: Princeton University Press.

Ludwig, David. 2016. Ontological Choices and the Value-Free Ideal. *Erkenntnis* 81 (6): 1253–1272.

Major, John S. 1993. *Heaven and Earth in Early Han Thought: Chapters Three, Four, and Five of the Huainanzi*. In *SUNY Studies in Chinese Philosophy and Culture*. Albany, NY: State University of New York Press.

Makdisi, George. 1970. Madrasa and University in the Middle Ages. *Studia Islamica* 32:255–264.

Makdisi, George. 1974. The Scholastic Method in Medieval Education: An Inquiry into its Origins in Law and Theology. *Speculum* 49 (4): 640–661.

Marques, Victor, and Carlos Brito. 2014. The Rise and Fall of the Machine Metaphor: Organizational Similarities and Differences between Machines and Living Beings. *Verifiche: Rivista trimestrale di scienze umane* 43 (1–3): 77–111.

Merton, Robert K. 1938. Science, Technology and Society in Seventeenth Century England. *Osiris* 4:360–632.

Merton, Robert K. 1942. Science and Technology in a Democratic Order. *Journal of Legal and Political Sociology* 1 (1): 115–126.

Milton, J. R. 1980. John Locke and the Nominalist Tradition. In *John Locke: Symposium Wolfenbüttel 1979*, ed. Reinhard Brandt, 128–145. Berlin: Walter de Gruyter & Co.

Murdoch, John. 1974. Philosophy and the Enterprise of Science in the Later Middle Ages. In *The Interaction Between Science and Philosophy*, The Van Leer Jerusalem Foundation Series, ed. Y. Elkana, 51–74. Atlantic Highlands, NJ: Humanities Press.

Nicholas of Cusa. 1996. *On Wisdom and Knowledge.* Translated by Jasper Hopkins. Minneapolis, MN: The Arthur J. Banning Press.

Nicholson, Daniel J. 2019. Is the Cell *Really* a Machine? *Journal of Theoretical Biology* 477:108–126.

Novikoff, Alex J. 2012. Toward a Cultural History of Scholastic Disputation. *The American Historical Review* 117 (2): 331–364.

Ornstein, Martha. 1928. *The Rôle of Scientific Societies in the Seventeenth Century.* Chicago: University of Chicago Press.

Pepper, Stephen. 1982. Metaphor in Philosophy. *The Journal of Mind and Behavior* 3 (3–4): 197–205.

Polanyi, Michael. 1963. The Potential Theory of Adsorption. *Science NS* 141 (3585): 1010–1013.

Popkin, Richard H. 2003. *The History of Scepticism: From Savonarola to Bayle. Revised and.* Expanded ed. Oxford: Oxford University Press.

Proctor, Robert N. 1991. *Value-Free Science: Purity and Power in Modern Knowledge.* Cambridge, MA: Harvard University Press.

Rattansi, P. M. 1968. The Intellectual Origins of the Royal Society. *Notes and Records of the Royal Society of London* 23 (2): 129–143.

Sabetti, Filippo. 2004. Local Roots of Constitutionalism. *Perspectives on Political Science* 33 (2): 70–78.

Sarao, K. T. S. 2017. Vihāra. In *Buddhism and Jainism*, The Encyclopedia of Indian Religions, ed. K. T. S. Sarao and Jeffery D. Long, 1328–1331. Dordrecht: Springer.

Scammell, G. V. 1969. The New Worlds and Europe in the Sixteenth Century. *The Historical Journal* 12 (3): 389–412.

Schüssler, Rudolf. 2009. Jean Gerson, Moral Certainty, and the Renaissance of Ancient Scepticism. *Renaissance Studies* 23 (4): 445–462.

Sextus, Empiricus. 1933. *Against the Logicians.* Translated by R. G. Bury. Loeb Classical Library. London: William Heinemann, Cambridge, MA: Harvard University Press.

Sextus, Empiricus. 1976. *Outlines of Pyrrhonism.* Translated by R. G. Bury, Loeb Classical Library. Cambridge. MA: Harvard University Press, London: William Heinemann.

Shapin, Steven, and Simon Schaffer. 1985. *Leviathan and the Air-Pump: Hobbes, Boyle, and the Experimental Life.* Princeton, NJ: Princeton University Press.

Smith, Pamela H. 2004. *The Body of the Artisan: Art and Experience in the Scientific Revolution*. Chicago: Chicago University Press.

Son, Bong-Ho. 1986. Facts and Values in Neo-Confucianism. In *Facts and Values: Philosophical Reflections from Western and Non-Western Perspectives*, Martinus Nijhoff Philosophy Library, ed. M. C. Doeser and I. N. Kraay, vol. 19, 149–159. Dordrecht: Kluwer.

Sutton, Clive. 1994. *Nullius in Verba* and *Nihil in Verbis*: Public Understanding of the Role of Language in Science. *British Journal for the History of Science* 27 (1): 55–64.

Swieżawski, Stefan. 2022. Beginnings of Modern Christian Aristotelianism. Translated by Grzegorz Czemiel and Tomasz Pałkowski. *Roczniki Filozoficzne* 70(4): 7–26.

van den Daele, Wolfgang. 1977. The Social Construction of Science: Institutionalisation and Definition of Positive Science in the Latter Half of the Seventeenth Century. In *The Social Production of Scientific Knowledge*, Sociology of the Sciences: A Yearbook, ed. Everett Mendelsohn, Peter Weingart, and Richard Whitley, vol. 1, 27–54. Dordrecht: D. Reidel.

Van Dyck, Maarten. 2022. Mechanical Philosophy: Science of Mechanics. In *Encyclopedia of Early Modern Philosophy and the Sciences*, ed. Dana Jalobeanu and Charles T. Wolfe, 1223–1233. Cham: Springer.

Wootton, David. 2015. *The Invention of Science: A New History of the Scientific Revolution*. New York: HarperCollins.

CHAPTER 5

Conclusion

Abstract What, then, is my answer to the question, 'Is science Western'? The thinkers I have called 'relativizers' are not entirely mistaken. Despite its ancient roots, modern science employs metatheoretical assumptions that have been influenced by the history of early modern Europe. These include a preference for mechanical metaphors, the practice of organized scepticism, and a clear distinction between facts and values. But these assumptions are open to revision in the light of new discoveries. On the operational level of scientific practice – the level at which discoveries are made – the sciences employ refined versions of epistemic practices that all humans share. Taking this fact into account, it seems the 'universalists' are correct: science is not merely Western; it is a set of practices without cultural boundaries.

Keywords Metatheoretical • Metaphors • Scepticism

Is science Western? Yes, and no. Modern science *is* 'Western' in a 'historical origins' sense of the term, insofar as it began to take on its present forms in early modern Europe. The historical context in which it emerged gave early modern science what appear to be distinctively Western characteristics: it created a preference for mechanical metaphors, gave an

G. W. Dawes, *Is Science Western?*,
https://doi.org/10.1007/978-3-032-10641-4_5

institutional context to the practice of organized scepticism, and encouraged a clear distinction between facts and values. I have described these as metatheoretical characteristics, which continue to shape scientific practice. But if modern science has some features that reflect its European origins, the practices it employs on what I have called its operational level are widely shared. The scientific revolution of the seventeenth century was not a *creatio e nihilo*. It involved a coming together of mathematical techniques and practices of experimentation developed in the ancient world and transmitted from other civilizations, particularly those of India and the Muslim world. More broadly, the sciences employ refined versions of epistemic practices found in all societies. These include observation-based reasoning, the practice of measurement, and the use of the core concepts that shape everyday knowledge.

An important idea here is that the operational level of scientific practice has priority over the theoretical and metatheoretical. The history of modern science shows that theories can be radically revised in the light of new data (as in the case of the move from Newtonian physics to relativity theory). But it also shows that science's metatheoretical assumptions—the assumptions that are most clearly 'Western' in origin—are open to challenge. The scientific community has not, thankfully, abandoned the practice of organized scepticism, which is arguably a key element in its success. But the explanations offered by modern physics are no longer 'mechanistic' in the sense favoured by seventeenth-century natural philosophers. Even in modern biology the time-honoured machine metaphor is under some threat. By embracing systems theory, some biologists are returning to a root metaphor that was familiar to Aristotle: that of the organism. Insofar as living organisms seek what is good for them, the metaphor of the organism may yet bring questions about value back into science.

At the beginning of this study I contrasted the views of the relativizers—those inclined to speak of modern science as 'Western science'—and the universalists—those who insist that scientific knowledge has no cultural boundaries. The present study lends support to the universalist view. If the operational level of science has priority over the metatheoretical, and if science, at its operational level, employs refined versions of practices found everywhere, there is no reason to regard scientific knowledge is limited to the societies in which it originally emerged.

There is much that scientists can learn from the forms of knowledge found in non-scientific cultures. Many of the peoples we call 'indigenous', for example, have a wealth of practical knowledge about their natural environment. The differing worldviews of non-Western societies can even help scientists revise their metatheoretical assumptions. While modern science has favoured the root metaphor of the machine, other societies have different roots metaphors—such as the Chinese idea of 'resonance' (*gǎnyìng*)—which may yet prove to be operationally fruitful. But while exposure to differing forms of knowledge may help scientists transcend the European origins of their disciplines, there is no reason to regard the conclusions of modern science as merely 'local'.

In arguing that modern science employs refined versions of epistemic practices found everywhere, I have highlighted concepts, practices, and capabilities that all human beings share. The idea that the human mind works in similar ways in all societies was once widely accepted. A popular term for this idea was 'the psychic unity of mankind'. It fell out of favour the mid-twentieth century, as anthropologists and historians highlighted the diversity of cultural worlds. But the idea is slowly coming back into favour, as cognitive science provides evidence that humans share a body of core cognitive capabilities. It does, however, remain controversial. Its advocates are sometimes accused of a kind of cultural assimilation, assuming that all peoples are like 'us', inhabitants of a world that is (by historical standards) WEIRD (Western, Educated, Industrialized, Rich and Democratic).

Cultural assimilation is clearly a danger. But in trying to avoid it, we should not fall into the error of thinking of cultures as incommensurable doxastic worlds. If there are instances of incommensurability between the ways of thinking found in different societies, they are merely local. Only subsets of our differing ways of representing the world are unable to be directly compared. Even apparently incommensurable features can be compared indirectly, for they derive from ways of knowing found in all human societies. The most important of these are the practices that constitute everyday practical knowledge, practices that scientific practice refines. Yes, modern science does have some culturally distinctive features. But it draws on ways of knowing found throughout human history and across differing societies. To this extent it is universal.

Index

G. W. Dawes, *Is Science Western?*,
https://doi.org/10.1007/978-3-032-10641-4

FSC
www.fsc.org
MIX
Papier | Fördert
gute Waldnutzung
FSC® C083411

Kolibri 360 GmbH
Ferdinand-Jühlke-Straße 7
99095 Erfurt, Deutschland
produktsicherheit@kolibri360.de